T0135549

A polarized discrete ordinate scattering model for radiative transfer simulations in spherical atmospheres with thermal source

Vom Fachbereich für Physik und Elektrotechnik
der Universität Bremen
zur Erlangung des akademischen Grades eines
Doktor der Naturwissenschaften (Dr. rer. nat.)
genehmigte Dissertation

von
Dipl. Phys. Claudia Emde

Dezember 2004

Berichte aus dem Institut für Umweltphysik – Band 25
herausgegeben von:

Dr. Georg Heygster
Universität Bremen, FB 1, Institut für Umweltphysik,
Postfach 33 04 40, D-28334 Bremen
URL http://www.iup.physik.uni-bremen.de
E-Mail iupsekr@uni-bremen.de

Die vorliegende Arbeit ist die inhaltlich unveränderte Fassung einer Dissertation, die im Dezember 2004 dem Fachbereich Physik/Elektrotechnik der Universität Bremen vorgelegt und von Prof. Dr. Klaus Künzi sowie Prof. Dr. Clemens Simmer begutachtet wurde. Das Promotionskolloquium fand am 28. Januar 2005 statt.

Bibliografische Information Der Deutschen Bibliothek
Die Deutsche Bibliothek verzeichnet diese Publikation in der Deutschen Nationalbibliografie; detaillierte bibliografische Daten sind im Internet über http://dnb.ddb.de abrufbar.

ISBN 3-8325-0855-4 ISSN 1615-6862

Logos Verlag Berlin
Comeniushof
Gubener Straße 47
D-10243 Berlin
Telefon (0 30) 42 85 10 90
URL http://www.logos-verlag.de

Layout: Lothar Meyer-Lerbs, Bremen

Contents

Abstract

This work describes the development of the new discrete ordinate scattering algorithm, which is a part of the Atmospheric Radiative Transfer Simulator (ARTS). Furthermore, applications of the algorithm, which was implemented to study for example the influence of cirrus clouds on microwave limb sounding, are presented.

The model development requires as a theoretical basis the electromagnetic scattering theory. The basic quantities are defined and different methods to compute single scattering properties of small particles are discussed. The phenomenological derivation of the vector radiative transfer equation, which is the basic equation of the model, is outlined. In order to represent clouds as scattering media in radiative transfer models, information about their micro-physical state is required as an input for calculating the scattering properties. The micro-physical state of a cloud is defined by the phase of the cloud particles, the particle size and shape distributions, the particle orientation, the ice mass or the liquid water content, and the temperature.

The model uses the Discrete Ordinate ITerative (DOIT) method to solve the vector radiative transfer equation. The implementation of a discrete ordinate method is challenging due to the spherical geometry of the model atmosphere, which is required for the simulation of limb radiances. The involved numerical issues, grid optimization and interpolation methods, are discussed.

The new scattering algorithm was compared to three other models, which were developed during the same time period as the DOIT algorithm. Overall, the agreement between the models was very good, giving confidence in new models.

Scattering simulations are presented for limb- and down-looking geometries, for one-dimensional and three-dimensional spherical atmospheres. They were performed for the frequency bands of the Millimeter Wave Acquisitions for Stratosphere/Troposphere Exchange Research (MASTER) instrument, and for selected frequencies of the Earth Observing System Microwave Limb Sounder (EOS MLS). The simulations show the impact of cloud particle size, shape and orientation on the brightness temperatures and on the polarization of microwave radiation in the atmosphere. The cloud effect is much larger for limb radiances than for nadir radiances. Particle size is a very important parameter in all of the simulations. The polarization signal is small for simulations with randomly oriented particles whereas for horizontally aligned particles with random azimuthal orientation the polarization signal is significant. Moreover, the effect of particle shape is only relevant for oriented cloud particles. The simulations show that it is essential to use a three-dimensional scattering model for inhomogeneous cloud layers.

Publications

The work described in this text has given rise to a number of publications:

Journal Articles

1. A detailed description of the DOIT scattering model including several examples is published in:

   Emde, C., S. A. Buehler, C. Davis, P. Eriksson, Sreerekha T. R. and C. Teichmann, A polarized discrete ordinate scattering model for simulations of limb and nadir longwave measurements in 1D/3D spherical atmospheres, *J. Geophys. Res.*, in press 2004.

2. First limb scattering simulations for MASTER frequency bands are presented in:

   Emde, C., S. A. Buehler, P. Eriksson and T. R. Sreerekha (2004), The effect of cirrus clouds on microwave limb radiances, *J. Atmos. Res., 72(1–4)*, 383–401, doi:10.1016/j.atmosres.2004.03.023.

3. The intercomparison of the ARTS-DOIT model with the radiative transfer model KOPRA is described in:

   Höpfner, M. and C. Emde (2005), Comparison of single and multiple scattering approaches for the simulation of limb-emission observations in the mid-IR, *J. Quant. Spectrosc. Radiat. Transfer, 91(3)*, 275–285, doi:10.1016/j.jqsrt.2004.05.066.

4. The following article includes a description of the Monte Carlo scattering model, which has been implemented in ARTS by C. Davis, and several calculations using this model:

   Davis, C., C. Emde and R. Harwood, A 3D polarized reversed Monte

Carlo radiative transfer model for mm and sub-mm passive remote sensing in cloudy atmospheres, *IEEE T. Geosci. Remote*, in press, 2004.

5. A prototype of the DOIT scattering module and a sensitivity study of cloud parameters on nadir radiances is presented in:

   Sreerekha, T. R., S. A. Buehler and C. Emde (2002), A simple new radiative transfer model for simulating the effect of cirrus clouds in the microwave spectral region, *J. Quant. Spectrosc. Radiat. Transfer*, 75, 611–624.

6. A paper including an explanation of polarization induced by spherical particles using the DOIT model has been submitted:

   Teichmann, C., S. A. Buehler and C. Emde Understanding the polarization signal of spherical particles for microwave limb radiances *J. Geophys. Res.*, submitted 2004.

7. The ARTS clear sky model was intercompared with several radiative transfer models. The results are published in the following paper:

   Melsheimer, C., C. Verdes, S. A. Buehler, C. Emde, P. Eriksson, D. G. Feist, S. Ichizawa, V. O. John, Y. Kasai, G. Kopp, N. Koulev, T. Kuhn, O. Lemke, S. Ochiai, F. Schreier, T. R. Sreerekha, M. Suzuki, C. Takahashi, S. Tsujimaru and J. Urban, Intercomparison of General Purpose Clear Sky Atmospheric Radiative Transfer Models for the Millimeter/Submillimeter Spectral Range, *Radio Sci.*, in press, 2004.

Technical Reports

8. The ARTS user guide also includes a detailed theoretical description and furthermore information about the usage of the model:

   Eriksson, P., S. A. Buehler, C. Emde, T. R. Sreerekha, C. Melsheimer and O. Lemke (2004), ARTS-1-1 User Guide, University of Bremen, 308 pages, regularly updated versions available at www.sat.uni-bremen.de/arts/.

9. A large part of the work was carried out in the context of ESCTEC studies: The *UTLS study* consisted of two tasks. The topic of task 1 was “2D retrievals of cloud-free scenes”. Task 2 was about 2D retrievals in the presence of clouds, for this part the DOIT model was developed and applied for simulations as well as for model comparisons. The executive summary includes the most important findings and conclusions:

   Kerridge, B., V. Jay, J. Reburn, R. Siddans, B. Latter, F. Lama, A. Dudhia, D. Grainger, A. Burgess, M. Höpfner, T. Steck, C. Emde, P. Eriksson, M. Ekström, A. Baran and M. Wickett (2004), Consideration of mission studying chemistry of the UTLS, Executive Summary, *ESTEC Contract No 15457/01/NL/MM.*

10. The results of task 2 are compiled in:

   Kerridge, B., V. Jay, J. Reburn, R. Siddans, B. Latter, F. Lama, A. Dudhia, D. Grainger, A. Burgess, M. Höpfner, T. Steck, G. Stiller, S. Buehler, C. Emde, P. Eriksson, M. Ekström, A. Baran and M. Wickett (2004), Consideration of mission studying chemistry of the UTLS, Task 2 Report, *ESTEC Contract No 15457/01/NL/MM.*

11. The final report includes the major results of task 1 and task 2:

   Kerridge, B., V. Jay, J. Reburn, R. Siddans, B. Latter, F. Lama, A. Dudhia, D. Grainger, A. Burgess, M. Höpfner, T. Steck, G. Stiller, S. Buehler, C. Emde, P. Eriksson, M. Ekström, A. Baran and M. Wickett (2004), Consideration of mission studying chemistry of the UTLS, Final Report, *ESTEC Contract No 15457/01/NL/MM.*

12. For the *RT study* a literature review of radiative transfer models for the microwave region which include scattering was performed:

   Claudia Emde and Sreerekha T. R. (2004), Development of a RT model for frequencies between 200 and 1000 GHz, WP1.2 Model Review, *ESTEC Contract No AO/1-4320/03/NL/FF.*

Articles in Conference Proceedings

13. Preliminary results obtained by using the DOIT scattering module are shown in:

    Emde, C., S. Buehler, Sreerekha T. R. (2003), Modeling polarized microwave radiation in a 3D spherical cloudy atmosphere, *In: Electromagnetic and Light Scattering – Theory and applications VII*, Edited by Wriedt, T., Universität Bremen, ISBN 3-88722-579-1.

14. Nadir simulations to estimate the cloud effect for AMSU-B channels are published in:

    Sreerekha, T. R., C. Emde and S. A. Buehler, Using a new Radiative Transfer Model to estimate the Effect of Cirrus Clouds on AMSU-B Radiances, *In: Twelfth International TOVS Study Conference (ITSC – XII)*, Lorne, Australia, February 2002.

15. The DOIT model was compared to the RTTOVSCAT model and to AMSU radiances. The results are presented in the conference proceedings:

    English, S. J., U. O'Keeffe, T. R. Sreerekha, S. A. Buehler, C. Emde and A. M. Doherty (2003), A Comparison of RTTOVSCATT with ARTS and AMSU Observations, Using the Met Office Mesoscale Model Short Range Forecasts of Cloud Ice and Liquid Water, In: *Thirteenth International TOVS Study Conference (ITSC – XIII)*, St. Adele, Montreal, Canada.

Preface

To improve existing climate models it is very important to extend the knowledge about cirrus cloud parameters, as such clouds cover more than 20% of the globe (Wang et al., 1996) and play an important role in the Earth's radiation budget (Arking, 1991). Depending on cloud altitude and micro-physical properties, clouds can either cause warming or cooling at the Earth's surface. So far clouds are not well treated in Global Climate Models (GCM) because of uncertainties concerning the properties of cirrus clouds and because of the complex interaction between radiation, micro-physics and dynamics in these clouds. Moreover it is essential to consider clouds for the evaluation of limb measurements of trace gases in the upper troposphere. Clouds, especially cirrus, with particle sizes exceeding microwave wavelengths, can severely disturb trace gas measurements. On the other hand it is possible to obtain cloud information from microwave limb radiances affected by cirrus clouds. This requires a radiative transfer model that can simulate the scattering effect of cirrus clouds.

In particular the effective radius R_{eff} of cloud particles is important for the radiative properties of clouds. For a given frequency, R_{eff} largely determines the relation between ice mass content (IMC) and cloud optical thickness (Evans et al., 1998). Parameterizations of R_{eff} have been retrieved from combined lidar and radar reflectivity (Donovan, 2003) or from observations made in situ using aircraft mounted instruments (e.g., Kinne et al., 1997; McFarquhar and Heymsfield, 1997). The Submillimeter-Wave Cloud Ice Radiometer (SWCIR) to fly on an aircraft has been developed to retrieve upper tropospheric IMC and R_{eff} (Evans et al., 2002).

Satellite remote sensing techniques in the thermal infrared can only be applied for thin cirrus clouds consisting of small ice particles as saturation is reached for moderate optical depths (Stubenrauch et al., 1999). Only ice particle properties of the uppermost cloud layers can be measured. Disadvantages of visible and near-infrared solar reflection methods include that they cannot measure low optical depth clouds over brighter land surfaces.

There are several studies about the sensitivity of cirrus clouds on microwave nadir radiances (e.g., Evans et al., 1998; Skofronik-Jackson et al., 2002). They show that the brightness temperature depression depends strongly on particle size and IMC. Passive nadir-viewing techniques cannot sufficiently resolve the vertical distribution of IMC. Millimeter-wave limb sounding is a well established technique for the observation of atmospheric trace gases in the stratosphere and upper troposphere. This technique can provide higher resolutions than nadir techniques for the same frequencies. Instruments using this technique are the Earth Observing System Microwave Limb Sounder (EOS MLS) (Waters et al., 1999), the Millimeter Atmospheric Sounder (MAS) (Hartmann et al., 1996) and the Millimeter Wave Acquisitions for Stratosphere/Troposphere Exchange Research (MASTER) instrument (Buehler, 1999). Recently, instruments have moved towards higher frequencies into the submillimeter-wave region, examples of this type of instrument are Odin-SMR (Murtagh et al., 2002) and the Superconduction Submillimeter-Wave Limb Emission Sounder (SMILES) (Buehler et al., 2005b).

A number of well established radiative transfer models exist for the clear sky case, notably the public domain Atmospheric Radiative Transfer Simulator (ARTS) (Buehler et al., 2005a), which was taken as the platform for the new scattering model described in this thesis. The model development is a challenging task for various reasons: Firstly, cloud coverage is vertically and horizontally inhomogeneous which implies that a three-dimensional (3D) model is unavoidable for the simulation of realistic cases. Especially for limb measurements, the 3D spherical geometry is required as the observed region in the atmosphere has a horizontally large extent. However, for largely extended thin cirrus clouds, it makes sense to use a one-dimensional (1D) model, because it can be much more efficient compared to a full 3D model. Secondly, cirrus clouds consist of particles of different sizes and shapes. Since particle scattering due to non-spherical particles leads to polarization effects (Czekala and Simmer, 1998), the vector radiative transfer equation (VRTE) has to be used in the model to obtain the full Stokes vector, not just the intensity of the radiation.

Liquid water clouds are not so problematic, because liquid water drops mainly act as absorbers, not as scatterers. Cirrus clouds, on the other hand, have a low absorption coefficient (see for example Mishchenko et al., 2002) and a rather large scattering coefficient. Aerosol scattering needs to be considered in the infra-red. Molecular Rayleigh scattering, though important for optical wavelength, can be neglected at microwave and infra-red wavelengths.

A survey of existing freely available radiative transfer models yielded none that were well-suited to the requirements described above. For instance the 3D Monte Carlo models described in Liu et al. (1996) and Roberti et al. (1994) are only applicable for 3D-cartesian atmospheres. The 3D discrete ordinate models SHDOM (Evans, 1998) and VDOM (Haferman et al., 1997) also assume a cartesian geometry. For this reason they are not applicable for limb simulations. Other discrete ordinate models, for example MWMOD (Simmer, 1993) and VDISORT (Schulz and Stamnes, 2000), use one-dimensional (1D) plane-parallel geometries. Another well known method is the Eddington approximation (e.g., Kummerow, 1993), which is also not well-suited to the limb sounding problem, as it is only valid in plane-parallel atmospheres. A simple 1D plane-parallel model using a prototype of the iterative solution method described in this thesis is presented in Sreerekha et al. (2002).

In the new version of ARTS two scattering methods have been implemented: a backward Monte Carlo Method (Davis et al., 2004) and the DOIT (Discrete Ordinate ITerative) (Emde et al., 2004a) method being presented in this work. Both methods work in 3D spherical atmospheres and both can simulate polarization effects due to aspherical particles. The DOIT method works also in 1D spherical atmospheres. The implementation of the DOIT method is very similar to discrete ordinate method (DOM) implementations for instance in SHDOM or VDOM. The originality of the DOIT method is, that the DOM has been adapted to a spherical geometry, which is essential for the simulation of limb radiances. The model can be applied in the microwave and in the infrared wavelength regions.

The present thesis consists of nine chapters. Chapter 1 gives an introduction to the theoretical concepts of the radiative transfer theory for scattering media. Basic quantities are defined and an outline of the phenomenological derivation of the vector radiative transfer equation is given. Chapter 2 introduces concepts and definitions of ARTS, which was used as a platform to implement the DOIT algorithm. Chapter 3 gives a brief overview of cloud micro-physics and it introduces different methods to calculate scattering properties of cloud particles. It is shown that the T-matrix method is the most appropriate to be used for the new scattering model.

The DOIT algorithm is described in detail in Chapter 4, which starts with the theoretical basis of discrete ordinates and afterwards explains the numerical optimizations, which were necessary for efficiency reasons. In Chapter 5 the 1D comparisons with the models FM2D and KOPRA are shown. Here ARTS-DOIT was used as a reference model, since it is the more general and more accurate model. Furthermore a 3D comparison with the ARTS Monte Carlo model is presented. Note that the two scattering models presented here are the first models which are able to simulate polarization in a 3D spherical atmosphere in the microwave region.

Chapter 6 presents 1D simulations for the MASTER instrument, where the effect of cloud parameters like effective radius and ice mass constant is investigated. First simulations using the full capabilities of the new model, i.e., polarization and 3D geometry, are shown in Chapter 7. In Chapter 8 simulations for the EOS MLS instrument are presented. Scattering and polarization of thermal radiation in a thin layer tropical cirrus cloud are investigated.

The final Chapter 9 consists of the overall summary and of conclusions.

1 Theoretical background

This chapter introduces the theoretical background which is essential to develop a radiative transfer model including scattering. The theory is based on concepts of electrodynamics, starting from the Maxwell equations. An elementary book for electrodynamics is written by Jackson (1998). For optics and scattering of radiation by small particles the reader may refer for instance to van de Hulst (1957) and Bohren and Huffman (1998). The notation used in this chapter is mostly adapted from the book "Scattering, Absorption, and Emission of Light by Small Particles" by Mishchenko et al. (2002). Several lengthy derivations of formulas, which are not shown in detail here, can also be found in this book. The purpose of this chapter is to provide definitions and give ideas, how these definitions can be derived using principles of electromagnetic theory. For the derivation of the radiative transfer equation an outline of the traditional phenomenological approach is given.

1.1 Basic definitions

From the Maxwell equations one can derive the formula for the electromagnetic field vector $\boldsymbol{E}$ of a plane electromagnetic wave propagating in a homogeneous medium without sources:

$$\boldsymbol{E}(\boldsymbol{r},t) = \boldsymbol{E}_0 \exp\left(-\frac{\omega}{c} m_{\mathrm{I}} \hat{\boldsymbol{n}} \cdot \boldsymbol{r}\right) \exp\left(\mathrm{i}\frac{\omega}{c} m_{\mathrm{R}} \hat{\boldsymbol{n}} \cdot \boldsymbol{r} - \mathrm{i}\omega t\right), \tag{1.1}$$

where $\boldsymbol{E}_0$ is the amplitude of the electromagnetic wave in vacuum, c is the speed of light in vacuum, ω is the angular frequency, $\boldsymbol{r}$ is the position vector and $\hat{\boldsymbol{n}}$ is a real unit vector in the direction of propagation. The complex refractive index m is

$$m = m_{\mathrm{R}} + \mathrm{i}m_{\mathrm{I}} = c\sqrt{\epsilon\mu}, \tag{1.2}$$

where m_R is the non-negative real part and m_I is the non-negative imaginary part. Furthermore μ is the permeability of the medium and ϵ the permittivity. For a vacuum, $m = m_R = 1$. The imaginary part of the refractive index, if it is non-zero, determines the decay of the amplitude of the wave as it propagates through the medium, which is thus absorbing. The real part determines the phase velocity $v = c/m_R$. The time-averaged Poynting vector $\boldsymbol{P}(\boldsymbol{r})$, which describes the flow of electromagnetic energy, is defined as

$$\boldsymbol{P}(\boldsymbol{r}) = \frac{1}{2}\mathrm{Re}\big(\langle \boldsymbol{E}(\boldsymbol{r})\rangle \times \langle \boldsymbol{H}^*(\boldsymbol{r})\rangle\big), \tag{1.3}$$

where $\boldsymbol{H}$ is the magnetic field vector and the $*$ denotes the complex conjugate. The Poynting vector for a homogeneous wave is given by

$$\langle \boldsymbol{P}(\boldsymbol{r})\rangle = \frac{1}{2}\mathrm{Re}\left(\sqrt{\frac{\epsilon}{\mu}}\right)|\boldsymbol{E}_0|^2 \exp\left(-2\frac{\omega}{c}m_I\hat{\boldsymbol{n}}\cdot\boldsymbol{r}\right)\hat{\boldsymbol{n}}. \tag{1.4}$$

Equation (1.4) shows that the energy flows in the direction of propagation and its absolute value $I(\boldsymbol{r}) = |\langle \boldsymbol{P}(\boldsymbol{r})\rangle|$, which is usually called intensity (or irradiance), is exponentially attenuated. Rewriting Equation (1.4) gives

$$I(\boldsymbol{r}) = I_0 \exp(-\alpha^p \hat{\boldsymbol{n}}\cdot\boldsymbol{r}), \tag{1.5}$$

where I_0 is the intensity for $\boldsymbol{r} = \boldsymbol{0}$. The absorption coefficient α^p is

$$\alpha^p = 2\frac{\omega}{c}m_I = \frac{4\pi m_I}{\lambda} = \frac{4\pi m_I \nu}{c}, \tag{1.6}$$

where λ is the free-space wavelength and ν the frequency. Intensity has the dimension of monochromatic flux [energy/(area $\times$ time)].

1.2 Definition of the Stokes parameters

Sensors usually do not measure directly the electric and the magnetic fields associated with a beam of radiation. They measure quantities that are time averages of real-valued linear combinations of products of field vector components and have the dimension of intensity. Examples of such observable quantities are the Stokes parameters. Figure 1.1 shows the coordinate system used to describe the direction of propagation $\hat{\boldsymbol{n}}$ and the polarization state of a plane electromagnetic wave.

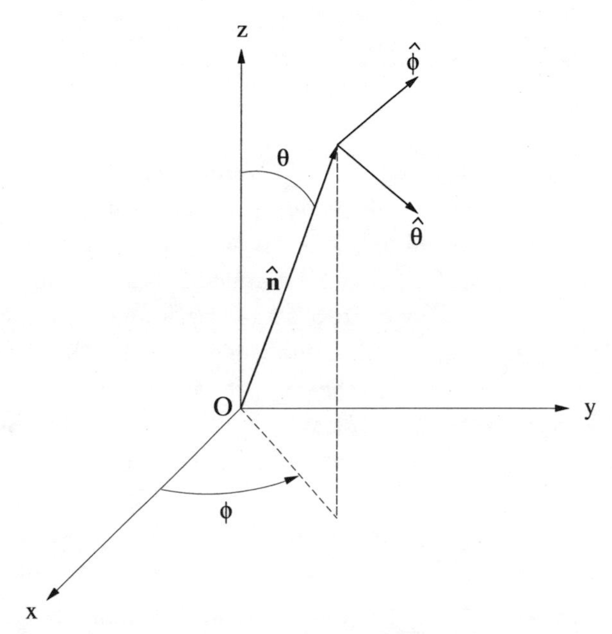

Figure 1.1: Coordinate system to describe the direction of propagation and the polarization state of a plane electromagnetic wave (adapted from Mishchenko).

The unit vector $\hat{\boldsymbol{n}}$ can equivalently be described by a couplet (θ, ϕ), where $\theta \in [0, \pi]$ is the polar (zenith) angle and $\phi \in [0, 2\pi)$ is the azimuth angle. The electric field at the observation point is given by $\boldsymbol{E} = \boldsymbol{E}_\theta + \boldsymbol{E}_\phi$, where $\boldsymbol{E}_\theta$ and $\boldsymbol{E}_\phi$ are the θ- and ϕ-components of the electric field vector. $\boldsymbol{E}_\theta$ lies in the meridional plane, which is the plane through $\hat{\boldsymbol{n}}$ and the z-axis, and $\boldsymbol{E}_\phi$ is perpendicular to this plane. $\boldsymbol{E}_\theta$ and $\boldsymbol{E}_\phi$ are often called $\boldsymbol{E}_\mathrm{v}$ and $\boldsymbol{E}_\mathrm{h}$ in the microwave remote sensing literature.

The Stokes parameters are defined as follows:

$$I = \frac{1}{2}\sqrt{\frac{\epsilon}{\mu}}(E_{\mathrm{v}}E_{\mathrm{v}}^{*} + E_{\mathrm{h}}E_{\mathrm{h}}^{*}), \tag{1.7}$$

$$Q = \frac{1}{2}\sqrt{\frac{\epsilon}{\mu}}(E_{\mathrm{v}}E_{\mathrm{v}}^{*} - E_{\mathrm{h}}E_{\mathrm{h}}^{*}), \tag{1.8}$$

$$U = -\frac{1}{2}\sqrt{\frac{\epsilon}{\mu}}(E_{\mathrm{v}}E_{\mathrm{h}}^{*} + E_{\mathrm{h}}E_{\mathrm{v}}^{*}), \tag{1.9}$$

$$V = i\frac{1}{2}\sqrt{\frac{\epsilon}{\mu}}(E_{\mathrm{h}}E_{\mathrm{v}}^{*} - E_{\mathrm{v}}E_{\mathrm{h}}^{*}). \tag{1.10}$$

They are commonly defined as a 4×1 column vector $\boldsymbol{I}$, which is known as the Stokes vector. Since the Stokes parameters are real-valued and have the dimension of intensity, they can be measured directly with suitable instruments. The Stokes parameters are a complete set of quantities needed to characterize a plane electromagnetic wave. They carry information of the complex amplitudes and the phase difference. The first Stokes parameter I is the intensity and the other components Q, U and V describe the polarization state of the wave. The Stokes parameters of a plane monochromatic wave are related by the quadratic identity

$$I^2 = Q^2 + U^2 + V^2. \tag{1.11}$$

The definition of a monochromatic plane wave implies that the complex amplitude $\boldsymbol{E_0}$ and the phase differences are constant. In the case of natural radiation the amplitudes and phases fluctuate, since the radiation originates from several sources that do not emit radiation coherently, and since the emission from one source usually has very short coherence times. This means that we usually have a superposition of radiation from several incoherent sources, and that the polarization state of the radiation from each source fluctuates as well. Such fluctuations have time scales that are longer than the period $(2\pi/\omega)$ of the oscillation, but that are still shorter than the integration time of the instrument that measures the radiation. Thus, the instrument measures an incoherent superposition of time averages over the fluctuating polarization. If the fluctuations are not completely random, the radiation is called partially polarized.

Since the different sources and/or emission events are assumed to be incoherent, the Stokes parameters can simply be added up:

$$I = \sum_i I_i, \quad Q = \sum_i Q_i, \quad U = \sum_i U_i, \quad V = \sum_i V_i. \tag{1.12}$$

The equality Equation (1.11) still holds for each contribution i, but for the resulting I, Q, U, V, we have in general the inequality

$$I^2 \geq Q^2 + U^2 + V^2. \tag{1.13}$$

The degree of polarization p is defined as

$$p = \frac{\sqrt{Q^2 + U^2 + V^2}}{I}. \tag{1.14}$$

For completely polarized radiation, $Q^2 + U^2 + V^2 = I^2$, thus $p = 1$, and for unpolarized radiation, $Q = U = V = 0$, thus $p = 0$.

In addition to the degree of polarization, p, the degree of linear polarization is defined as

$$p_{\text{lin}} = \frac{\sqrt{Q^2 + U^2}}{I}, \tag{1.15}$$

and the the degree of circular polarization is defined as

$$p_{\text{circ}} = \frac{V}{I}. \tag{1.16}$$

1.3 Scattering, absorption and thermal emission by a single particle

A parallel monochromatic beam of electromagnetic radiation propagates in vacuum without any change in its intensity or polarization state. A small particle, which is interposed into the beam, can cause several effects:

Absorption: The particle converts some of the energy contained in the beam into other forms of energy.

Elastic scattering: Part of the incident energy is extracted from the beam and scattered into all spatial directions at the frequency of the incident beam. Scattering can change the polarization state of the radiation.

Extinction: The energy of the incident beam is reduced by an amount equal to the sum of absorption and scattering.

Dichroism: The change of the polarization state of the beam as it passes a particle.

Thermal emission: If the temperature of the particle is non-zero, the particle emits radiation in all directions over a large frequency range.

The beam is an oscillating plane magnetic wave, whereas the particle can be described as an aggregation of a large number of discrete elementary electric charges. The incident wave excites the charges to oscillate with the same frequency and thereby radiate secondary electromagnetic waves. The superposition of these waves gives the total elastically scattered field.

One can also describe the particle as an object with a refractive index different from that of the surrounding medium. The presence of such an object changes the electromagnetic field that would otherwise exist in an unbounded homogeneous space. The difference of the total field in the presence of the object can be thought of as the field *scattered* by the object. The angular distribution and the polarization of the scattered field depend on the characteristics of the incident field as well as on the properties of the object as its size relative to the wavelength and its shape, composition and orientation.

1.3.1 Definition of the amplitude matrix

For the derivation of a relation between the incident and the scattered electric field we consider a finite scattering object in the form of a single body or a fixed aggregate embedded in an infinite homogeneous, isotropic and non-absorbing medium. We assume that the individual bodies forming the scattering object are sufficiently large that they can be characterized by optical constants appropriate to bulk matter, not to optical constants appropriate for single atoms or molecules. Solving the Maxwell equations for the internal volume, which is the interior of the scattering object, and the external volume one can derive a formula, which expresses the total electric field everywhere in space in terms of the incident field and the field inside the scattering object. Applying the far field approximation gives a relation between incident and scattered field, which is that of a spherical wave. The amplitude matrix $\boldsymbol{S}(\hat{\boldsymbol{n}}^{\text{sca}}, \hat{\boldsymbol{n}}^{\text{inc}})$ includes this relation:

$$\begin{pmatrix} E_\theta^{\text{sca}}(r\hat{\boldsymbol{n}}^{\text{sca}}) \\ E_\phi^{\text{sca}}(r\hat{\boldsymbol{n}}^{\text{sca}}) \end{pmatrix} = \frac{e^{ikr}}{r} \boldsymbol{S}(\hat{\boldsymbol{n}}^{\text{sca}}, \hat{\boldsymbol{n}}^{\text{inc}}) \begin{pmatrix} E_{0\theta}^{\text{inc}} \\ E_{0\phi}^{\text{inc}} \end{pmatrix}. \tag{1.17}$$

The amplitude matrix depends on the directions of incident $\hat{\boldsymbol{n}}^{\text{inc}}$ and scattering $\hat{\boldsymbol{n}}^{\text{sca}}$ as well as on size, morphology, composition, and orientation of the scattering object with respect to the coordinate system. The distance between the origin and the observation point is denoted by r and the wave number of the external volume is denoted by k.

The amplitude matrix provides a complete description of the scattering pattern in the far field zone. The amplitude matrix explicitly depends on ϕ^{inc} and ϕ^{sca} even when θ^{inc} and/or θ^{sca} equal 0 or π.

1.3.2 Phase matrix

The phase matrix $\boldsymbol{Z}$ describes the transformation of the Stokes vector of the incident wave into that of the scattered wave for scattering directions away from the incidence direction ($\hat{\boldsymbol{n}}^{\text{sca}} \neq \hat{\boldsymbol{n}}^{\text{inc}}$),

$$\boldsymbol{I}^{\text{sca}}(r\hat{\boldsymbol{n}}^{\text{sca}}) = \frac{1}{r^2} \boldsymbol{Z}(\hat{\boldsymbol{n}}^{\text{sca}}, \hat{\boldsymbol{n}}^{\text{inc}}) \boldsymbol{I}^{\text{inc}}. \tag{1.18}$$

The 4×4 phase matrix can be written in terms of the amplitude matrix elements for single particles (Mishchenko et al., 2002). All elements of the phase matrix have the dimension of area and are real. As the amplitude matrix, the phase matrix depends on ϕ^{inc} and ϕ^{sca} even when θ^{inc} and/or θ^{sca} equal 0 or π. In general, all 16 elements of the phase matrix are non-zero, but they can be expressed in terms of only seven independent real numbers. Four elements result from the moduli $|S_{ij}|$ $(i, j = 1, 2)$ and three from the phase-differences between S_{ij}. If the incident beam is unpolarized, i.e., $\boldsymbol{I}^{\text{inc}} = (I^{\text{inc}}, 0, 0, 0)^T$, the scattered light generally has at least one non-zero Stokes parameter other than intensity:

$$I^{\text{sca}} = Z_{11} I^{\text{inc}}, \tag{1.19}$$

$$Q^{\text{sca}} = Z_{21} I^{\text{inc}}, \tag{1.20}$$

$$U^{\text{sca}} = Z_{31} I^{\text{inc}}, \tag{1.21}$$

$$V^{\text{sca}} = Z_{41} I^{\text{inc}}. \tag{1.22}$$

This is the phenomena is traditionally called "polarization". The non-zero degree of polarization Equation (1.14) can be written in terms of the phase matrix elements

$$p = \frac{\sqrt{Z_{21}^2 + Z_{31}^2 + Z_{41}^2}}{Z_{11}}. \tag{1.23}$$

1.3.3 Extinction matrix

In the special case of the exact forward direction ($\hat{\boldsymbol{n}}^{\text{sca}} = \hat{\boldsymbol{n}}^{\text{inc}}$) the attenuation of the incoming radiation is described by the extinction matrix $\boldsymbol{K}$. In terms of the Stokes vector we get

$$\boldsymbol{I}(r\hat{\boldsymbol{n}}^{\text{inc}})\Delta S = \boldsymbol{I}^{\text{inc}}\Delta S - \boldsymbol{K}(\hat{\boldsymbol{n}}^{\text{inc}})\boldsymbol{I}^{\text{inc}} + O(r^{-2}). \tag{1.24}$$

Here ΔS is a surface element normal to $\hat{\boldsymbol{n}}^{\text{inc}}$. The extinction matrix can also be expressed explicitly in terms of the amplitude matrix. It has only seven independent elements. Again the elements depend on ϕ^{inc} and ϕ^{sca} even when the incident wave propagates along the z-axis.

1.3.4 Absorption vector

The particle also emits radiation if its temperature T is above zero Kelvin. According to Kirchhoff's law of radiation the emissivity equals the absorptivity of a medium under thermodynamic equilibrium. The energetic and polarization characteristics of the emitted radiation are described by a four-component Stokes emission column vector $\boldsymbol{a}(\hat{\boldsymbol{r}}, T, \omega)$. The emission vector is defined in such a way that the net rate, at which the emitted energy crosses a surface element ΔS normal to $\hat{\boldsymbol{r}}$ at distance r from the particle at frequencies from ω to $\omega + \Delta\omega$, is

$$W^e = \frac{1}{r^2}\boldsymbol{a}(\hat{\boldsymbol{r}}, T, \omega)B(T, \omega)\Delta S \Delta\omega, \tag{1.25}$$

where W^e is the power of the emitted radiation and B is the Planck function. In order to calculate $\boldsymbol{a}$ we assume that the particle is placed inside an opaque cavity of dimensions large compared to the particle and any wavelengths under consideration. We have thermodynamic equilibrium if the cavity and the particle are maintained at the constant temperature T. The emitted radiation inside the cavity is isotropic, homogeneous, and unpolarized. We can represent this radiation as a collection of quasi-monochromatic, unpolarized, incoherent beams propagating in all directions characterized by the Planck blackbody radiation

$$B(T, \omega)\Delta S \Delta\Omega = \frac{\hbar\omega^3}{2\pi^2 c^2 \left[\exp\left(\frac{\hbar\omega}{k_B T}\right) - 1\right]}\Delta S\, \Delta\Omega, \tag{1.26}$$

where $\Delta\Omega$ is a small solid angle about any direction, $\hbar$ is the Planck constant divided by 2π, and k_B is the Boltzmann constant. The blackbody Stokes vector is

$$\boldsymbol{I}_b(T, \omega) = \begin{pmatrix} B(T, \omega) \\ 0 \\ 0 \\ 0 \end{pmatrix}. \tag{1.27}$$

For the Stokes emission vector, which we also call particle absorption vector, we can derive

$$a_i^p(\hat{\boldsymbol{r}}, T, \omega) = K_{i1}(\hat{\boldsymbol{r}}, \omega) - \int_{4\pi} \mathrm{d}\hat{\boldsymbol{r}}' Z_{i1}(\hat{\boldsymbol{r}}, \hat{\boldsymbol{r}}', \omega), \quad i = 1, \ldots, 4. \tag{1.28}$$

This relation is a property of the particle only, and it is valid for any particle, in thermodynamic equilibrium or non-equilibrium.

1.3.5 Optical cross sections

The optical cross-sections are defined as follows: The product of the scattering cross section C_{sca} and the incident monochromatic energy flux gives the total monochromatic power removed from the incident wave as a result of scattering into all directions. The product of the absorption cross section C_{abs} and the incident monochromatic energy flux gives the power which is removed from the incident wave by absorption. The extinction cross section C_{ext} is the sum of scattering and absorption cross section. One can express the extinction cross sections in terms of extinction matrix elements

$$C_{\text{ext}} = \frac{1}{I^{\text{inc}}} \big(K_{11}(\hat{\boldsymbol{n}}^{\text{inc}})I^{\text{inc}} + K_{12}(\hat{\boldsymbol{n}}^{\text{inc}})Q^{\text{inc}} + K_{13}(\hat{\boldsymbol{n}}^{\text{inc}})U^{\text{inc}} + K_{14}(\hat{\boldsymbol{n}}^{\text{inc}})V^{\text{inc}}\big), \tag{1.29}$$

and the scattering cross section in terms of phase matrix elements

$$C_{\text{sca}} = \frac{1}{I^{\text{inc}}} \int_{4\pi} \mathrm{d}\hat{\boldsymbol{r}} \big(Z_{11}(\hat{\boldsymbol{r}}, \hat{\boldsymbol{n}}^{\text{inc}})I^{\text{inc}} + Z_{12}(\hat{\boldsymbol{r}}, \hat{\boldsymbol{n}}^{\text{inc}})Q^{\text{inc}} + Z_{13}(\hat{\boldsymbol{r}}, \hat{\boldsymbol{n}}^{\text{inc}})U^{\text{inc}} + Z_{14}(\hat{\boldsymbol{r}}, \hat{\boldsymbol{n}}^{\text{inc}})V^{\text{inc}}\big). \tag{1.30}$$

The absorption cross section is the difference between extinction and scattering cross section:

$$C_{\text{abs}} = C_{\text{ext}} - C_{\text{sca}}. \tag{1.31}$$

The single scattering albedo ω_0, which is a commonly used quantity in radiative transfer theory, is defined as the ratio of the scattering and the extinction cross section:

$$\omega_0 = \frac{C_{\text{sca}}}{C_{\text{ext}}} \leq 1. \tag{1.32}$$

All cross sections are real-valued positive quantities and have the dimension of area.

The phase function is generally defined as

$$p(\hat{\boldsymbol{r}}, \hat{\boldsymbol{n}}^{\text{inc}}) = \frac{4\pi}{C_{\text{sca}} I^{\text{inc}}} \big(Z_{11}(\hat{\boldsymbol{r}}, \hat{\boldsymbol{n}}^{\text{inc}})I^{\text{inc}} + Z_{12}(\hat{\boldsymbol{r}}, \hat{\boldsymbol{n}}^{\text{inc}})Q^{\text{inc}} + Z_{13}(\hat{\boldsymbol{r}}, \hat{\boldsymbol{n}}^{\text{inc}})U^{\text{inc}} + Z_{14}(\hat{\boldsymbol{r}}, \hat{\boldsymbol{n}}^{\text{inc}})V^{\text{inc}}\big). \tag{1.33}$$

The phase function is dimensionless and normalized:

$$\frac{1}{4\pi} \int_{4\pi} p(\hat{\boldsymbol{r}}, \hat{\boldsymbol{n}}^{\text{inc}})\, \mathrm{d}\hat{\boldsymbol{r}} = 1. \tag{1.34}$$

1.4 Scattering, absorption and emission by ensembles of independent particles

The formalism described in the previous chapter applies only for radiation scattered by a single body or a fixed cluster consisting of a limited number of components. In reality, one normally finds situations, where radiation is scattered by a very large group of particles forming a constantly varying spatial configuration. Clouds of ice crystals or water droplets are a good example for such a situation. A particle collection can be treated at each given moment as a fixed cluster, but as a measurement takes a finite amount of time, one measures a statistical average over a large number of different cluster realizations.

Solving the Maxwell equations for a whole cluster, like a collection of particles in a cloud, is computationally too expensive. Fortunately, particles forming a random group can often be considered as independent scatterers. This approximation is valid under the following assumptions:

1. Each particle is in the far-field zone of all other particles.
2. Scattering by the individual particles is incoherent.

As a consequence of assumption 2, the Stokes parameters of the partial waves can be added without regard to the phase. If the particle number density is sufficiently small, the single scattering approximation can be applied. The scattered field in this approach is obtained by summing up the fields generated by the individual particles in response to the external field in isolation from all other particles. If the particle positions are random, one can show, that the phase matrix, the extinction matrix and the absorption vector are obtained by summing up the respective characteristics of all constituent particles.

1.4.1 Single scattering approximation

We consider a volume element containing N particles. We assume that N is sufficiently small, so that the mean distance between the particles is much larger than the incident wavelength and the average particle size. Furthermore we assume that the contribution of the total scattered signal of radiation scattered more than once is negligibly small. This is equivalent to the requirement

$$\frac{N\left\langle C_{\mathrm{sca}}\right\rangle}{l^2} \ll 1, \tag{1.35}$$

where $\langle C_{\mathrm{sca}}\rangle$ is the average scattering cross section per particle and l is the linear dimension of the volume element. The electric field scattered by the volume element can be written as the vector sum of the partial scattered fields scattered by the individual particles:

$$\boldsymbol{E}^{\mathrm{sca}}(\boldsymbol{r}) = \sum_{n=1}^{N} \boldsymbol{E_n}^{\mathrm{sca}}(\boldsymbol{r}). \tag{1.36}$$

As we assume single scattering the partial scattered fields are given according to Equation (1.17):

$$\begin{pmatrix} [E_n^{\mathrm{sca}}(\boldsymbol{r})]_\theta \\ [E_n^{\mathrm{sca}}(\boldsymbol{r})]_\phi \end{pmatrix} = \frac{e^{ikr}}{r} \boldsymbol{S}(\hat{\boldsymbol{r}}, \hat{\boldsymbol{n}}^{\mathrm{inc}}) \begin{pmatrix} E_{0\theta}^{\mathrm{inc}} \\ E_{0\phi}^{\mathrm{inc}} \end{pmatrix}, \tag{1.37}$$

where $\boldsymbol{S}$ is the total amplitude scattering matrix given by:

$$\boldsymbol{S}(\hat{\boldsymbol{r}}, \hat{\boldsymbol{n}}^{\mathrm{inc}}) = \sum_{n=1}^{N} e^{i\Delta_n} \boldsymbol{S}_n(\hat{\boldsymbol{r}}, \hat{\boldsymbol{n}}^{\mathrm{inc}}). \tag{1.38}$$

$\boldsymbol{S}_n(\hat{\boldsymbol{r}}, \hat{\boldsymbol{n}}^{\mathrm{inc}})$ are the individual amplitude matrices and the phase Δ_n is given by

$$\Delta_n = k\boldsymbol{r}_{\mathrm{On}} \cdot (\hat{\boldsymbol{n}}^{\mathrm{inc}} - \hat{\boldsymbol{r}}), \tag{1.39}$$

where the vector $\boldsymbol{r}_{\mathrm{On}}$ connects the origin of the volume element O with the nth particle origin (see Figure 1.2). Since Δ_n vanishes in forward direction and the individual extinction matrices can be written in terms of the individual amplitude matrix elements, the total extinction matrix is given by

$$\boldsymbol{K} = \sum_{n=1}^{N} \boldsymbol{K}_n = N\left\langle \boldsymbol{K} \right\rangle, \tag{1.40}$$

where $\langle \boldsymbol{K} \rangle$ is the average extinction matrix per particle. One can derive the analog equation for the phase matrix

$$\boldsymbol{Z} = \sum_{n=1}^{N} \boldsymbol{Z}_n = N\left\langle \boldsymbol{Z} \right\rangle, \tag{1.41}$$

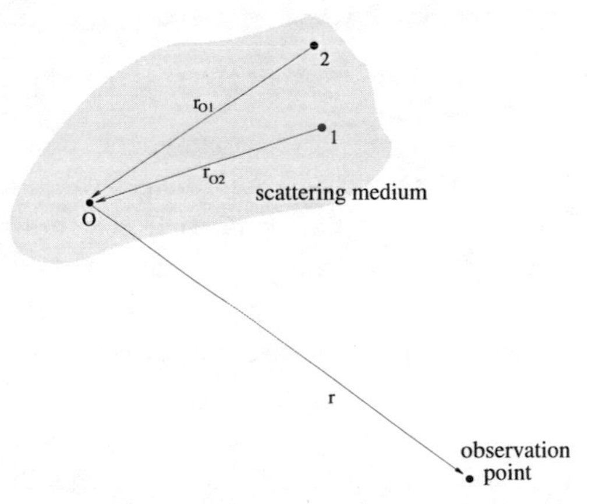

Figure 1.2: A volume element of a scattering medium conststing of a particle ensemble. O is the origin of the volume element, r_{O1} connects the origin with particle 1 and r_{O2} with particle 2. The observation point is assumed to be in the far-field zone of the volume element.

where $\langle \boldsymbol{Z} \rangle$ is the average phase matrix per particle. In almost all practical situations, radiation scattered by a collection of independent particles is incoherent, as a minimal displacement of a particle or a slight change in the scattering geometry changes the phase differences entirely. It is important to note, that the ensemble averaged phase matrix and the ensemble averaged extinction matrix have in general 16 independent elements. The relations between the matrix elements, which can be derived for single particles, do not hold for particle ensembles.

1.5 Phenomenological derivation of the radiative transfer equation

When the scattering medium contains a very large number of particles the single scattering approximation is no longer valid. In this case we have to take into account that each particle scatters radiation that has already been scattered by another particle. This means that the radiation leaving the medium has a significant multiple scattered component. The observation point is assumed to be in the far-field zone of each particle, but it is not necessarily in the far-field zone of the scattering medium as a whole. A traditional method in this case is to solve the radiative transfer equation. This approach still assumes, that the particles forming the scattering medium are randomly positioned and widely separated and that the extinction and the phase matrices of each volume element can be obtained by incoherently adding the respective characteristics of the constituent particles. In other words the scattering media is assumed to consist of a large number of discrete, sparsely and randomly distributed particles and is treated as continuous and locally homogeneous. Radiative transfer theory is originally a phenomenological approach based on considering the transport of energy through a medium filled with a large number of particles and ensuring energy conservation. Mishchenko (2002) has demonstrated that it can be derived from electromagnetic theory of multiple wave scattering in discrete random media under certain simplifying assumptions.

In the phenomenological radiative transfer theory, the concept of single scattering by individual particles is replaced by the assumption of scattering by a small homogeneous volume element. It is furthermore assumed that the result of scattering is not the transformation of a plane incident wave into a spherical scattered wave, but the transformation of the specific intensity vector, which includes the Stokes vectors from all waves contributing to the electromagnetic radiation field.

The vector radiative transfer equation (VRTE) is

$$\begin{aligned}\frac{\mathrm{d}\boldsymbol{I}(\boldsymbol{n},\nu)}{\mathrm{d}s} = &- \langle \boldsymbol{K}(\boldsymbol{n},\nu,T)\rangle \boldsymbol{I}(\boldsymbol{n},\nu) + \langle \boldsymbol{a}(\boldsymbol{n},\nu,T)\rangle B(\nu,T) \\ &+ \int_{4\pi} \mathrm{d}\boldsymbol{n}' \langle \boldsymbol{Z}(\boldsymbol{n},\boldsymbol{n}',\nu,T)\rangle \boldsymbol{I}(\boldsymbol{n}',\nu),\end{aligned} \tag{1.42}$$

where $\boldsymbol{I}$ is the specific intensity vector, $\langle \boldsymbol{K}\rangle$ is the ensemble-averaged extinction matrix, $\langle \boldsymbol{a}\rangle$ is the ensemble-averaged absorption vector, B is the Planck function and $\langle \boldsymbol{Z}\rangle$ is the ensemble-averaged phase matrix. Furthermore ν is the frequency of the radiation, T is the temperature, $\mathrm{d}s$ is a path-length-element of the propagation path and $\boldsymbol{n}$ the propagation direction. Equation (1.42) is valid for monochromatic or quasi-monochromatic radiative transfer. We can use this equation for simulating microwave radiative transfer through the atmosphere, as the scattering events do not change the frequency of the radiation.

The four-component specific intensity vector $\boldsymbol{I} = (I, Q, U, V)^T$ fully describes the radiation and it can directly be associated with the measurements carried out by a radiometer used for remote sensing. For the definition of the components of the specific intensity vector refer to Section 1.2, where the Stokes components are described. Since the specific intensity vector is a superposition of Stokes vectors, the polarization state of the specific intensity vector can be analysed in the same way as the polarization state of the Stokes vector.

The three terms on the right hand side of Equation (1.42) describe physical processes in an atmosphere containing different particle types and different trace gases. The first term represents the extinction of radiation traveling through the scattering medium. It is determined by the ensemble averaged extinction coefficient matrix $\langle \boldsymbol{K}\rangle$. For microwave radiation in cloudy atmospheres, extinction is caused by gaseous absorption, particle absorption and particle scattering. Therefore $\langle \boldsymbol{K}\rangle$ can be written as a sum of two matrices, the particle extinction matrix $\langle \boldsymbol{K}^p\rangle$ and the gaseous extinction matrix $\langle \boldsymbol{K}^g\rangle$:

$$\langle \boldsymbol{K}(\boldsymbol{n},\nu,T)\rangle = \langle \boldsymbol{K}^p(\boldsymbol{n},\nu,T)\rangle + \langle \boldsymbol{K}^g(\boldsymbol{n},\nu,T)\rangle. \tag{1.43}$$

The particle extinction matrix is the sum over the individual specific extinction matrices $\langle \boldsymbol{K}_i^p(\boldsymbol{n},\nu,T)\rangle$ of the N different particles types contained in the scattering medium weighted by their particle number densities n_i^p:

$$\langle \boldsymbol{K}^p(\boldsymbol{n},\nu,T)\rangle = \sum_{i=1}^{N} n_i^p \langle \boldsymbol{K}_i^p(\boldsymbol{n},\nu,T)\rangle . \tag{1.44}$$

A particle distribution, which can include various particle sizes, shapes and orientations, can be represented by a single particle type, since it is possible to derive an ensemble averaged phase matrix $\langle \boldsymbol{Z}_i\rangle$, an ensemble averaged extinction matrix $\langle \boldsymbol{K}_i\rangle$ and an ensemble averaged absorption vector $\langle \boldsymbol{a}_i\rangle$. The gaseous extinction matrix is directly derived from the scalar gas absorption. As there is no polarization due to gas absorption at cloud altitudes, the off-diagonal elements of the gaseous extinction matrix are zero. At very high altitudes above approximately 40 km there is polarization due to the Zeeman effect, mainly due to oxygen molecules. However, in the toposphere and stratosphere molecular scattering can be neglected in the microwave frequency range. Hence the coefficients on the diagonal correspond to the gas absorption coefficient:

$$\left\langle \boldsymbol{K}_{l,m}^g(\nu,T)\right\rangle = \begin{cases} \langle \alpha^g(\nu,T)\rangle & \text{if } l=m \\ 0 & \text{if } l\neq m. \end{cases} \tag{1.45}$$

where T is the temperature of the atmosphere and $\langle \alpha^g\rangle$ is the total scalar gas absorption coefficient, which is calculated from the individual absorption coefficients of all M trace gases $\alpha_i^g(P,\nu,T)$ and their volume mixing ratios n_i^g as:

$$\langle \alpha^g(\nu,T)\rangle = \sum_{i=1}^{M} n_i^g \alpha_i^g(\nu,T). \tag{1.46}$$

The second term in Equation (1.42) is the thermal source term. It describes thermal emission by gases and particles in the atmosphere. The ensemble averaged absorption vector $\langle \boldsymbol{a}\rangle$ is

$$\langle \boldsymbol{a}(\boldsymbol{n},\nu,T)\rangle = \langle \boldsymbol{a}^p(\boldsymbol{n},\nu,T)\rangle + \langle \boldsymbol{a}^g(\nu,T)\rangle , \tag{1.47}$$

where $\langle \boldsymbol{a}^p\rangle$ and $\langle \boldsymbol{a}^g\rangle$ are the particle absorption vector and the gas absorption vector, respectively. The particle absorption vector is a sum over the individual absorption vectors $\langle \boldsymbol{a}_i^p \rangle$, again weighted with n_i^p:

$$\langle \boldsymbol{a}^p(\boldsymbol{n}, \nu, T) \rangle = \sum_{i=1}^{N} n_i^p \langle \boldsymbol{a}_i^p(\boldsymbol{n}, \nu, T) \rangle . \tag{1.48}$$

The gas absorption vector is simply

$$\langle \boldsymbol{a}^g(\nu, T) \rangle = (\langle \alpha^p(\nu, T) \rangle, 0, 0, 0)^T . \tag{1.49}$$

The last term in Equation (1.42) is the scattering source term. It adds the amount of radiation which is scattered from all directions $\boldsymbol{n}'$ into the propagation direction $\boldsymbol{n}$. The ensemble averaged phase matrix $\langle \boldsymbol{Z} \rangle$ is the sum of the individual phase matrices $\langle \boldsymbol{Z}_i \rangle$ weighted with n_i^p:

$$\langle \boldsymbol{Z}(\boldsymbol{n}, \boldsymbol{n}', \nu, T) \rangle = \sum_{i=1}^{N} n_i^p \langle \boldsymbol{Z}_i(\boldsymbol{n}, \boldsymbol{n}', \nu, T) \rangle . \tag{1.50}$$

The scalar radiative transfer equation (SRTE)

$$\begin{aligned} \frac{\mathrm{d}I}{\mathrm{d}s}(\boldsymbol{n}, \nu) = &- \langle K_{11}(\boldsymbol{n}, \nu, T) \rangle I(\boldsymbol{n}, \nu) + \langle a_1(\boldsymbol{n}, \nu, T) \rangle B(\nu, T) \\ &+ \int_{4\pi} \mathrm{d}\boldsymbol{n}' \langle Z_{11}(\boldsymbol{n}, \boldsymbol{n}', \nu, T) \rangle I(\boldsymbol{n}', \nu) \end{aligned} \tag{1.51}$$

can be used presuming that the radiation field is unpolarized. This approximation is reasonable if the scattering medium consists of spherical or completely randomly oriented particles, where $\langle \boldsymbol{K}^p \rangle$ is diagonal and only the first element of $\langle \boldsymbol{a}^p \rangle$ is non-zero.

2 ARTS – the atmospheric radiative transfer system

This chapter introduces basic concepts and definitions of the ARTS model. It provides a brief summary of functions and methods used for the scattering simulations. Many of the functions, for example functions for the calculation of propagation paths, could be shared between the clear sky part of the model and the scattering part.

2.1 History

A lot of effort has been put in the development of dedicated forward models for different sensors. All of these models have many parts in common. While appropriate for operational data analysis, such specialized models are not appropriate for scientific studies of new sensor concepts, since they can not easily be adapted to new instruments. This was the reason for the development of more general forward models like the program FORWARD (Eriksson and Buehler, 2001), which was mostly written by J. Langen in the time period 1991–1998 at the University of Bremen, or the Skuld model mainly developed by Eriksson et al. (2002) during 1997–1998. Although these models were rather general and have been used successfully over the years, both suffered from being not easily modifiable and extendable. This has lead to the development of a model which emphasizes modularity, extendibility, and generality.

It was decided that the development work should be shared between the Bremen and Chalmers universities, with Bremen being largely responsible for the overall program architecture and the absorption part, Chalmers being largely responsible for the radiative transfer part and the calculation of Jacobians. The project was put under a GNU general public license (Stallman, 2002), in order to give the right legal framework for such a true collaboration.

The program, along with extensive documentation, is freely available on the Internet, under http://www.sat.uni-bremen.de/arts/. The stable 1-0-x branch of the program is described in Buehler et al. (2005a). Stable means that there will be only bug fixes, no additions of new features.A great part of the work for this thesis is dedicated to the development of the new branch, 1-1-x, which can handle scattering in the atmosphere. The development team has been joined by C. Davis from the University of Edinburgh, who has used the ARTS model as a platform for the implementation of a Monte Carlo scattering module additionally to the discrete ordinate scattering module, which is presented in this thesis.

2.2 Definition of the atmosphere

2.2.1 Atmospheric dimensionality

The modeled atmosphere can be selected to have different dimensionalities:

3D This is the most general case, where the atmospheric fields vary in all three spatial coordinates. A spherical coordinate system is used where the dimensions are pressure (P), latitude (α) and longitude (β). Choosing this option allows to simulate realistic radiation fields, including strongly horizontally inhomogeneous cloud coverage.

1D A "1D" atmosphere is a spherically symmetric atmosphere, which means that atmospheric fields and the ground extend in all three dimensions, but they do not have a variation in latitude and longitude. Atmospheric fields for instance vary only as a function of altitude. The surface of the earth corresponds to a sphere. The 1D geometry is a crude approximation for scattering simulations, as a spherically symmetric cloud corresponds to a globally complete cloud coverage. This extreme case can be used to study the effect of scattering by largely extended thin cirrus clouds.

2D A 2D atmosphere extends inside a plane. A polar coordinate system, consisting of a radial and an angular coordinate, is used. The 2D case is most likely used for satellite measurements where the atmosphere is observed inside the orbit plane. Scattering calculations can not be performed in 2D geometry as there is no case involving clouds that give rise to a radiation field that fits into a 2D framework.

2.2.2 The cloud box

In order to save computational time, scattering calculations are limited to the part of the atmosphere containing clouds and other scattering objects. The atmospheric region in which scattering shall be considered is denoted as the cloud box. The cloud box is defined to be rectangular in the used coordinate system, with limits exactly at points of the involved grids. This means, for example, that the vertical limits of the cloud box are two pressure surfaces.

When defining the cloud box limits, one must avoid that radiation emerging from the cloud box reenters the cloud box at another point, because the scattering calculation takes as boundary condition the incoming clear sky field. If there is a large amount of ground reflection and if the optical depth under the cloud box is small, there should not be a gap between the ground and the cloud box. In this case the ground is included as a scattering object. However, for many cases it can be accepted to have a gap between the ground and the cloud box, with the gain that the cloud box can be made smaller. Such a case is when the ground is treated to act as blackbody, the ground is then not reflecting any radiation. Reflections from the ground can also be neglected if the zenith optical thickness of the atmosphere between the ground and cloud box is sufficiently high.

Figure 2.1 shows schematically 1D and 3D model atmospheres including a cloud box.

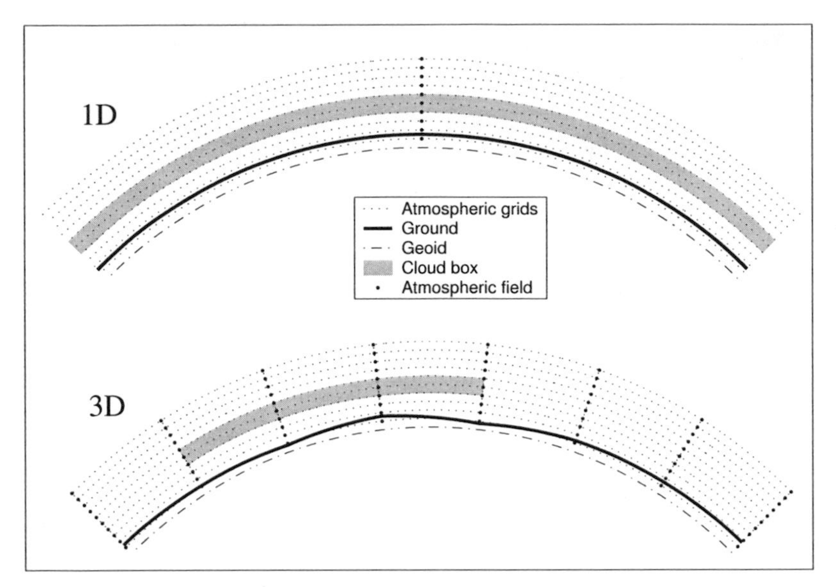

Figure 2.1: Top panel: Schematic of a 1D atmosphere. The atmosphere is here spherically symmetric. This means that the radius of the geoid, the ground and all atmospheric profiles are constant around the globe. The grey area indicates the cloud box. Bottom panel: Cross section of a 3D atmosphere. Atmospheric fields are defined on all grid points. The cloud box has a finite horizontal extent and the surface is not spherically symmetric.

2.3 Radiative transfer calculations

The radiative transfer (RT) calculations are divided into two separate parts, a clear sky part and the scattering calculations inside the cloud box. These parts have been implemented as two main modules with a well defined interface. The task of the scattering part is to determine the outgoing intensity field of the cloud box. The scattering calculations can be performed in any way as long as the outgoing field is provided. The outgoing field is then used as the radiative background for observation directions giving a propagation path that intersects with the cloud box.

The aim, when designing the clear sky and scattering modules, was that as many components as possible should be common. This is advantageous for many reasons, e.g., it decreases the amount of code to maintain, it facilitates detection of bugs, and enhances the consistency between the modules. An example of a common component is the calculation of propagation paths, where the same function for propagation path calculations is used in both modules.

The clear sky radiative transfer calculation is performed for a full measurement sequence. This means that the function calculates spectra for all positions of the sensor, and all pencil beam directions needed for the weighting with the antenna pattern. The inclusion of sensor characteristics etc. are not discussed here, details are given in Eriksson et al. (2004).

The clear sky part is vectorized in frequency (all monochromatic frequencies are handled in parallel), being in contrast to the scattering part. The number of Stokes components to consider can be set to any value from 1 to 4 (this is also valid for the scattering part). Polarized calculations (number of Stokes components > 1) can be performed independently from the cloud box being activated or not.

The RT calculations for a single pencil beam direction can be separated into three sub-tasks:

- Calculation of the propagation path.
- Determining the radiative background.
- Solving the radiative transfer equation.

2.3.1 Propagation paths

Any combination of sensor position and line-of-sight that makes sense with respect to the model atmosphere is allowed. The main restriction is that propagation paths are only allowed to enter or exit the model atmosphere at the top. This means that the propagation path can not exit the model atmosphere at a latitude end face for a 2D and 3D case.

Propagation paths are calculated backwards from the sensor to the practical starting point. If the sensor is placed outside the model atmosphere, geometrical calculations are used to find the exit point at the top of the atmosphere. Inside the model atmosphere, the path is calculated in steps, from one crossing of a grid cell boundary to the next

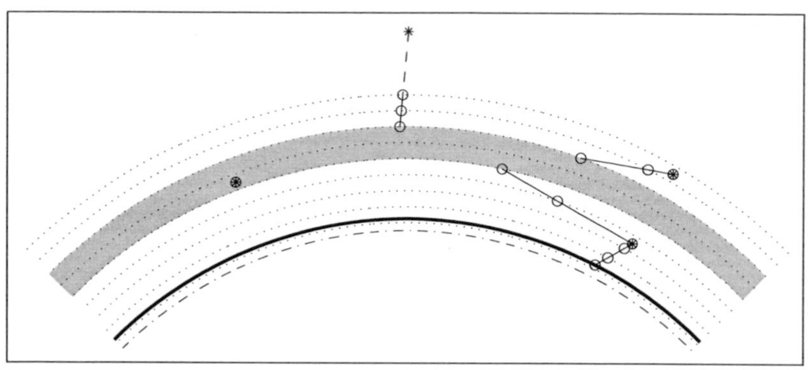

Figure 2.2: Propagation path examples for a 1D atmosphere. The dotted lines are the atmospheric grid, the dashed-dotted line is the geoid, the thick solid line is the ground and the cylindrical segment drawn with a thin solid line is the cloud box.

crossing. The functions to calculate such propagation path steps can be used both in the clear sky and scattering parts. Propagation paths are followed backwards until the top of the atmosphere, the ground or the cloud box (if activated) is reached. The propagation paths are described by a number of points along the path. Points are always included for crossings with the grids, tangent points (if such exist) and the position of the sensor, if placed inside the atmosphere. Depending on the function selected for the path step calculations, other points can be included, for instance to fulfill a criterion on the maximum length along the path between the points. Examples of propagation paths are given in Figure 2.2.

2.3.2 Radiative background

The radiative intensities at the starting point of the propagation path are denoted as the radiative background. Four possible radiative backgrounds exist:

1. Cosmic background when the propagation path starts at the top of the atmosphere.
2. The up-welling radiation from the ground when the propagation

path intersects with the ground. Emission and scattering properties of the ground need to be defined to determine the radiative background.
3. If the propagation path hits the surface of the cloud box, the radiative background is obtained by interpolating the radiation field leaving the cloud box. This interpolation considers the propagation direction of the path at the crossing point.
4. The internal intensity field of the cloud box is the radiative background for cases when the sensor is placed inside the cloud box.

2.3.3 Clear sky radiative transfer

The intensity matrix (holding all frequencies and Stokes components) is set to equal the radiative background. The calculations are then performed by solving the radiative transfer problem from one point of the propagation path to next, until the end point is reached.

The clear sky vector radiative transfer equation follows from the general VRTE Equation (1.42) by omitting the scattering integral and particle contributions to the extinction matrix and the absorption vector:

$$\frac{\mathrm{d}\boldsymbol{I}}{\mathrm{d}s}(\boldsymbol{n},\nu,T) = -\langle \boldsymbol{K}^g(\boldsymbol{n},\nu,T)\rangle \boldsymbol{I}(\boldsymbol{n},\nu,T) + \langle \boldsymbol{a}^g(\boldsymbol{n},\nu,T)\rangle B(\nu,T). \tag{2.1}$$

This equation can be solved analytically for constant coefficients. The extinction matrix $\langle \boldsymbol{K}^g(\boldsymbol{n},\nu,T)\rangle$ and the absorption vector $\langle \boldsymbol{a}^g(\boldsymbol{n},\nu,T)\rangle$ are averaged for one propagation path step. The averaging procedure will be described more detailed in Section 4.1. The solution is found using a matrix exponential approach (see Appendix B.1):

$$\boldsymbol{I}^i = e^{-\overline{\langle \boldsymbol{K}^g\rangle} s} \cdot \boldsymbol{I}^{i-1} + (\mathbb{I} - e^{-\overline{\langle \boldsymbol{K}^g\rangle} s})\overline{\langle \boldsymbol{K}^g\rangle}^{-1}(\overline{\langle \boldsymbol{a}^g\rangle}\,\bar{B}), \tag{2.2}$$

where $\overline{\langle \boldsymbol{K}^g\rangle}$ and $\overline{\langle \boldsymbol{a}^g\rangle}$ are the averaged quantities and i denotes a point in the propagation path.

2.4 Scattering

As mentioned above the task of the scattering module is to determine the outgoing radiation field on the boundary of the cloud box. This requires numerical methods to solve the VRTE (1.42) inside the cloud box as there is no analytical solution to the VRTE without any approximations. Two different approaches are implemented in ARTS: A backward Monte Carlo scheme which is briefly described in Section 5.3 and the discrete ordinate iterative approach, which has been developed by the author of this thesis and will be described in detail in Chapter 4. Several studies in which the DOIT method has been applied will be presented in the chapters 6 to 8.

2.5 Gas absorption

Calculating gas absorption in a line-by-line way is expensive, as sometimes contributions from thousands or ten thousands of spectral lines have to be taken into account. This needs to be done over and over again for each point in the atmosphere. The gas absorption coefficient does not depend directly on the position, but on the atmospheric state variables: pressure, temperature and trace gas concentrations. The basic idea in ARTS is to pre-calculate absorption for discrete combinations of these variables, store the values in a lookup table, and then interpolate them for the actual atmospheric state. The gas absorption coefficients are taken from spectral line catalog, for example from the HITRAN catalogs (Rothman et al., 1998).

2.6 Definition of clouds and atmospheric fields

In the Earth's atmosphere we find liquid water clouds consisting of approximately spherical water droplets and cirrus clouds consisting of ice particles of diverse shapes and sizes. We also find different kinds of aerosols. In order to take into account this variety, the model allows to define several *particle types*. A particle type is either a specified particle or a specified particle distribution, for example a particle ensemble following a gamma size distribution. The particles can be completely randomly oriented, azimuthally randomly oriented or arbitrarily oriented. For each particle type being a part of the modeled cloud field, a data file containing the single scattering properties ($\langle \boldsymbol{K}_i \rangle$, $\langle \boldsymbol{a}_i \rangle$, and $\langle \boldsymbol{Z}_i \rangle$), and the appropriate particle number density field is required. The particle number density fields are stored in data files, which include the field stored in a three-dimensional tensor and also the appropriate atmospheric grids (pressure, latitude and longitude grid). For each grid point in the cloud box the single scattering properties are averaged using the particle number density fields. In the scattering database the single scattering properties are not always stored in the same coordinate system. For instance for randomly oriented particles it makes sense to store the single scattering properties in the so-called scattering frame in order to reduce memory requirements (refer to Section 3.4 for more details).

The atmospheric fields, which are temperature, altitude, and volume mixing ratio fields, are stored in the same format as the particle number density fields.

2.7 Unit conventions

Internally the ARTS model uses SI[1] units for all quantities. However, SI units can sometimes be inconvenient, for example to represent the radiation field in the atmosphere. Therefore it is possible in ARTS to convert radiances from the SI unit [W s m^{-2} sr^{-1}] into a brightness temperature [K] unit. There are two brightness temperature (BT) definitions which can be applied:

1. **Planck BT:**

   The simplest case of remote sensing of temperature occurs when atmospheric extinction can be neglected. Then a satellite would just 'see' the thermal emission of the Earth's surface. One obtains the

1 SI units – Système International d'Unités

temperature of the surface from the measured radiance by inverting the Planck function Equation (1.26):

$$T_{\mathrm{Planck}}(I_b) = \frac{h\nu}{k_B \ln\left(\frac{2h\nu^3}{c^2 I_b} + 1\right)}. \tag{2.3}$$

More generally we can define a brightness temperature in terms of the radiance I_b using Equation (2.3), even in presence of extinction.

2. **Rayleigh Jeans BT:**
The definition of Planck BT can not be used for all Stokes components, because it is only defined for positive values of I_b. Since the Stokes components Q, U and V can be negative, we would like to have a conversion, which can also be applied for negative values. Another problem is the non-linearity of the Planck BT definition. The Stokes component Q is the difference between the vertically and the horizontally polarized parts of the radiation. Using the Planck BT definition, the value of Q would depend on whether we first transform I_v and I_h and take the difference of the obtained Planck BT, or we transform Q directly to Planck BT. The Rayleigh Jeans definition of BT is linear. The proportionality factor is derived from the Rayleigh Jeans approximation: At small frequencies ($h\nu \ll k_b T$) the Planck function is approximately

$$I_b(T, \omega) = \frac{2k_B\nu^2}{c^2} T. \tag{2.4}$$

Inverting Equation (2.4) yields the definition of Rayleigh Jeans BT:

$$T_{RJ}(I_b) = \frac{c^2}{2k_B\nu^2} I_b. \tag{2.5}$$

Since Rayleigh Jeans BT's can be used for all Stokes components, this BT definition is the only one used in this work. Note: Rayleigh Jeans BT are not equal to Planck BT, they are two different units to represent radiances.

3 Description of clouds as scattering media

This chapter deals with the representation of clouds as scattering media in radiative transfer models. An overview of cloud microphysics provides realistic ranges of particle sizes, shapes and ice mass contents of cirrus clouds. Different methods to calculate scattering properties for small particles are introduced. It is shown that the Rayleigh and Mie approximations are not sufficient for modeling radiative transfer through cirrus clouds. The cloud particles are not sufficiently small to be treated as Rayleigh scatterers. They are usually aspherical and often horizontally aligned, which makes it impossible to use the Mie theory, which is valid only for spherical particles. Although it can only handle rotationally symmetric particles, the T-matrix method was chosen to be used in ARTS, since it yields a rather good approximation for most realistic particles and it is widely used and tested. This chapter also introduces particle size distributions which are used for simulations in later chapters.

3.1 Microphysics of clouds

The earth's atmosphere consists of various particles: aerosols, water droplets, ice crystals, raindrops, snowflakes, graupel and hailstones. Cloud particles are the most important scatterers in the microwave region.

Clouds, which are composed of water droplets or ice crystals, are conventionally classified in terms of their position and appearance in the atmosphere. At mid-latitudes, clouds with base heights of about 6 km are defined as high clouds or cirrus clouds. The group of low clouds below about 2 km include stratus and cumulus. Middle clouds, between the high and the low clouds, include altocumulus and altostratus. The dispersion of particle sizes and their phase (liquid or ice) determines the microphysical state of a cloud. According to Liou (2002) low clouds and some middle clouds are generally composed of spherical water droplets with sizes ranging from 1 μm to 20 μm. The typical size for a water droplet is 5 μm. Middle clouds with temperatures warmer than about −20°C can contain super-cooled water droplets that coexist with ice particles. The small water droplets are spherical due to the surface tension. Larger raindrops deviate from the spherical shape while they are falling down. Cirrus clouds and some of the top and middle clouds contain ice crystals. The ice crystal shapes are irregular and depend on temperature, relative humidity and on the dynamics in the clouds, i.e., whether they undergo collision and coalescence processes. For humidities close to water saturation, the particles have prismatic skeleton shapes that occur in hollow and cluster crystals. These particles are referred to as bullet rosettes and they occur for example in cirrocumulus clouds. In cirrostratus clouds, where the relative humidity is close to ice saturation, ice crystals are predominantly individual and have shapes like columns, prisms, and plates. Between water and ice saturation, the ice crystals grow in the form of prisms.

Figure 3.1 shows a spectrum of ice crystal sizes and shapes as a function of height, relative humidity, and temperature in a typical midlatitude cirrus. Since ice crystal shape and size vary greatly with time and space, it is difficult to find representative values for remote sensing applications. Figure 3.2 shows five measured size distributions. The data is taken from Heymsfield and Platt (1984) and the figure is adapted from Liou (2002). The mean effective ice crystal size ranges from 10 μm to 124 μm.

In cirrus clouds ice particles are generally not randomly oriented. Laboratory experiments have shown, that cylinders (aspect ratio[1] < 1) tend to fall with their long axes horizontally oriented. Based on observations, columnar and plate crystals (aspect ratio > 1) tend to

1 The aspect ratio of a particle is its diameter divided by its length.

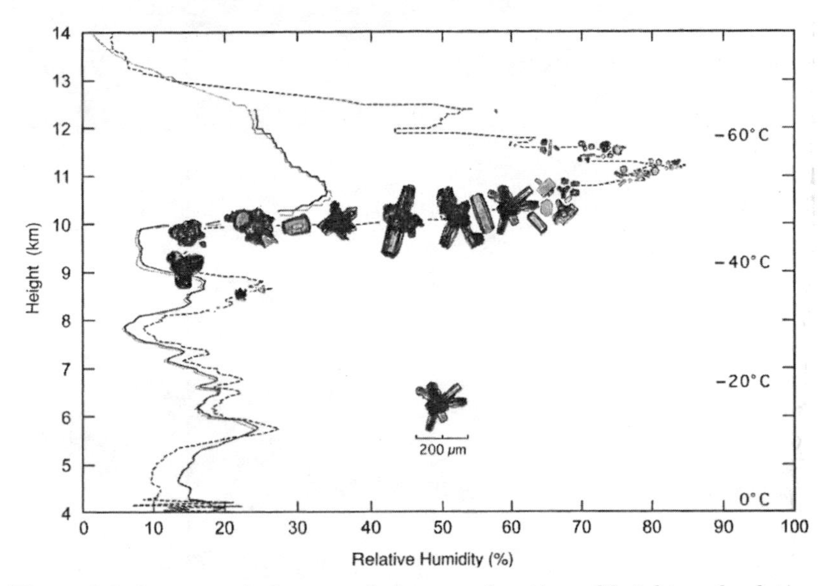

Figure 3.1: Ice crystal shape and size as a function of height and relative humidity captured by a replicator ballon sounding system in Marshall, Colorado on November 10, 1994. The relative humidity was measured by a cryogenic hygrometer (dashed line) and Vaisala RS80 instruments (solid line and dots). Figure adapted from Liou (2002).

fall with their major axes horizontally oriented. The orientation of ice particles in cirrus clouds has been observed by lidar measurements based on the depolarization technique in the backscattering direction (Liou, 2002). The measurements have shown, that specific orientations occur when the particles have relatively large sizes and well defined shapes, like cylinders or plates. If the ice crystals are irregular, such as aggregates, there is no preferred orientation. Moreover, smaller ice particles in cirrus clouds tend to be not randomly oriented in three-dimensional space if there is a substantial turbulence in the cloud. It has also been observed that ice particle orientation and alignment are strongly modulated by the electric field in clouds.

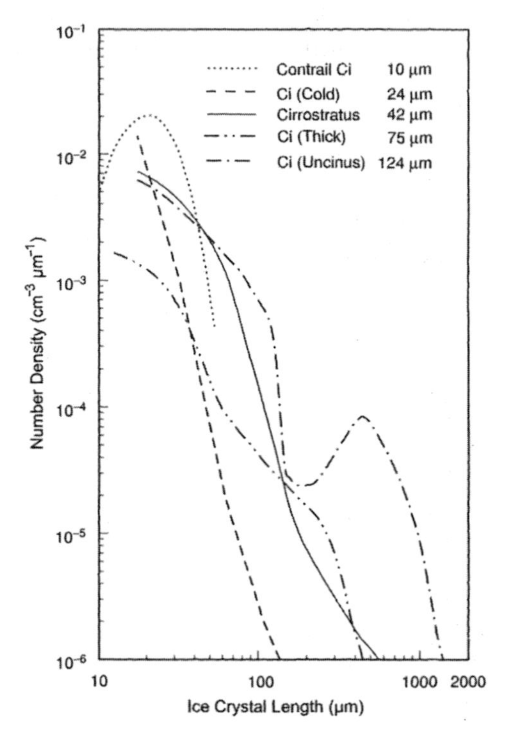

Figure 3.2: Ice crystal size distributions for midlatitude cirrus clouds covering a range of mean effective ice crystal sizes from 10 μm (Contrail), 24 μm (Cold), 42 μm (Cirrostratus), 75 μm (Thick), to 124 μm (Uncinus). The data was taken from Heymsfield and Platt (1984) and from Liou et al. (1998). Figure adapted from Liou (2002).

3.2 Coordinate systems: The laboratory frame and the scattering frame

For radiative transfer calculations we need a coordinate system to describe the direction of propagation. For this purpose we use the laboratory frame, which has been introduced in Chapter 1, Figure 1.1. The z-axis corresponds to the local zenith direction and the x-axis points towards the north-pole. The propagation direction is described by the local zenith angle θ and the local azimuth angle ϕ. This coordinate

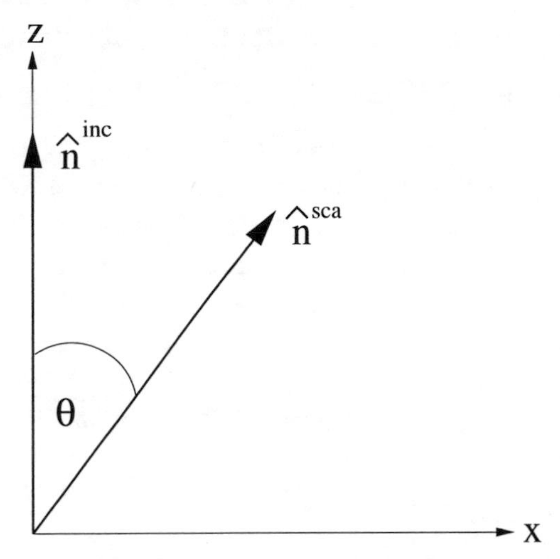

Figure 3.3: Illustration of the scattering frame. The z-axis coincides with the incident direction $\hat{\boldsymbol{n}}^{\rm inc}$. The scattering angle Θ is the angle between $\hat{\boldsymbol{n}}^{\rm inc}$ and $\hat{\boldsymbol{n}}^{\rm sca}$.

system is the most appropriate frame to describe the propagation direction and the polarization state of the radiation. However, in order to describe scattering of radiation by a particle or a particle ensemble, it makes sense to define another coordinate system taking into consideration the symmetries of the particle or the scattering medium, as one gets much simpler expressions for the single scattering properties. For macroscopically isotropic and mirror-symmetric scattering media it is convenient to use the scattering frame, in which the incidence direction is parallel to the z-axis and the x-axis coincides with the scattering plane, that is, the plane through the unit vectors $\hat{\mathbf{n}}^{\rm inc}$ and $\hat{\mathbf{n}}^{\rm sca}$. The scattering frame is illustrated in Figure 3.3. For symmetry reasons the single scattering properties defined with respect to the scattering frame can only depend on the scattering angle Θ,

$$\Theta = \arccos(\hat{\boldsymbol{n}}^{\rm inc} \cdot \hat{\boldsymbol{n}}^{\rm sca}), \tag{3.1}$$

between the incident and the scattering direction.

3.3 Methods to calculate scattering by small particles

This section introduces basic concepts, which are relevant for atmospheric scattering. It will be shown that neither the Rayleigh scattering approximation nor the Lorentz-Mie theory for scattering of radiation by small particles is generally appropriate to describe scattering of microwave radiation by cirrus cloud particles. The T-matrix method, one of the more elaborate methods, leads to a better understanding, especially of polarization due to cloud scattering.

An important quantity in scattering theory for small particles is the size parameter

$$x = \frac{2\pi r}{\lambda} = \frac{2\pi r \nu}{c}, \tag{3.2}$$

where r is the particle radius, λ is the wavelength and ν is the frequency of the incident radiation. The size of non-spherical particles is here defined by their equal volume sphere radius. The volume of a cylindrical particle for instance, which has an equal volume sphere radius of 75 μm, is identical to the volume of a sphere with a radius of 75 μm. Rayleigh scattering, the most simple theory, is valid for $x \ll 1$, i.e., if the particle size is much smaller than the wavelength of the incident radiation. If the wavelength is comparable to the particle size ($x \approx 1$), one can apply the Lorentz-Mie theory for spherical particles or the T-matrix method for spherical and non-spherical particles. For size parameters much greater than one, the geometrical optics approximation can be applied. Figure 3.4 shows size parameters as a function of frequency for different particle sizes of cloud ice particles. Only for very small cloud particles (10 μm) or frequencies below 100 GHz x is small enough such that Rayleigh scattering applies. The geometrical optics approximation can not be applied for any of these particles.

3.3.1 Rayleigh scattering

Rayleigh scattering is mostly applied for molecular scattering in the visible wavelength region or for the scattering of very low-frequency

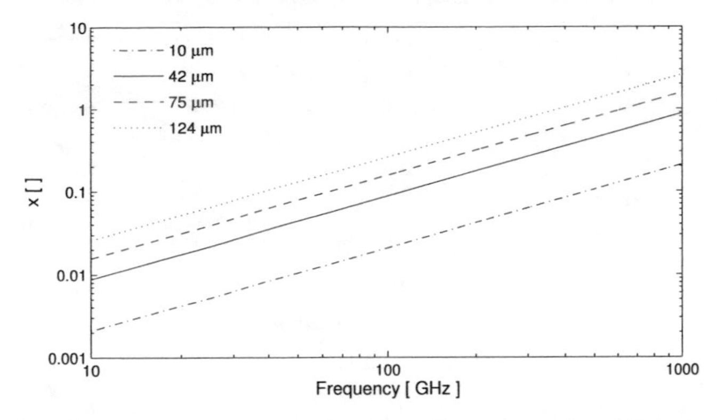

Figure 3.4: Size parameter x as a function of frequency for different particle sizes typical for cloud particles.

microwave radiation by hydrometeors. As air molecules are of several orders smaller than microwave wavelengths, molecular scattering can be neglected in the microwave wavelength region. Nevertheless, some aerosol or cloud particles can be sufficiently small to be treated as Rayleigh scatterers. The classical electro-dynamical solution yields the relation between incident and scattered intensity, which are denoted by I_0 and I respectively,

$$I = \frac{I_0}{r_p^2}\alpha^2\left(\frac{2\pi}{\lambda}\right)^4\frac{1+\cos^2\Theta}{2}, \tag{3.3}$$

where α is the polarizability of the small particle, r_p is the distance from the particle and Θ is the scattering angle. This is the formula derived by Rayleigh (1871). The formula shows that the intensity of sunlight scattered by a molecule is proportional to the incident intensity and inversely proportional to the square of the distance between the molecule and the observation point. It is also inversely proportional to λ^4 which explains the blue color of the sky. Since blue light is scattered more than red light, the sky, when viewed away from the sun disk appears blue. The Rayleigh phase function $p(\Theta)$ is given by

$$p(\Theta) = \frac{3}{4}(1+\cos^2\Theta). \tag{3.4}$$

The phase function is symmetric about the minimum at a scattering angle of 90°. An equal amount of radiation is scattered into the forward direction ($\Theta = 0°$) and into the backward direction ($\Theta = 180°$). The degree of linear polarization p_{lin} (Equation (1.15)) can also be derived for particles which are very small compared to the wavelength:

$$p_{\text{lin}} = \frac{\sin^2 \Theta}{\cos^2 \Theta + 1}. \tag{3.5}$$

In the forward and backward directions the scattered radiation remains completely unpolarized, whereas at a scattering angle of 90°, the scattered radiation becomes completely polarized. In other directions the radiation is partially polarized.

3.3.2 Lorentz-Mie theory for scattering by spherical particles

Starting from the Maxwell equations, one can derive analytically the single scattering properties of spherical particles in the far field approximation. For a detailed derivation refer, for example, to Liou et al. (1998) or Bohren and Huffman (1998). The MATLAB code developed by Mätzler (2002) was used to calculate phase functions for different particle sizes existing in clouds, corresponding to the effective radii of the size distributions shown in Figure 3.2. The particles were assumed to consist of ice. The refractive index of ice is calculated according to Mätzler (1998). The phase functions are computed for 89 GHz and 318 GHz and presented in Figure 3.5. For 89 GHz all phase functions are very close to the Rayleigh phase function, which is the expected result, as the size parameter for this frequency is always smaller than approximately 0.01. However, for 318 GHz the phase functions are very different from the Rayleigh phase function. Only for a particle radius of 10 µm it is still possible to use the Rayleigh approximation. The larger particles scatter more radiation into the forward and less into the backward direction. The minimum of the Rayleigh phase function is at a scattering angle of 90°. It shifts towards larger scattering angles as the size parameter increases.

Figure 3.6 shows the extinction efficiency Q_{ext}, the scattering efficiency Q_{sca} and the absorption efficiency Q_{abs} of spherical ice particles in the whole microwave-wavelength region. The efficiencies correspond

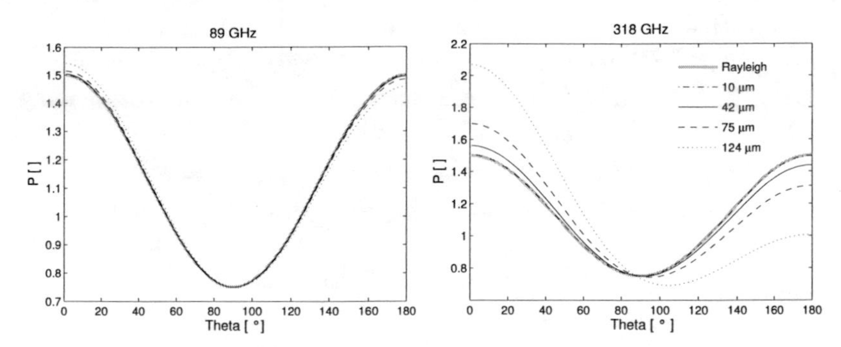

Figure 3.5: Grey lines: Rayleigh phase function. Black lines: Mie phase functions for different particle sizes. The left plot is for 89 GHz and the right plot for 318 GHz.

to the area normalized optical cross sections C_{ext}, C_{sca} and C_{abs}, which have been defined in Section 1.3.5, thus

$$Q_{\text{ext}} = \frac{C_{\text{ext}}}{\pi r^2}, \qquad Q_{\text{sca}} = \frac{C_{\text{sca}}}{\pi r^2}, \qquad Q_{\text{abs}} = \frac{C_{\text{abs}}}{\pi r^2}, \tag{3.6}$$

where r is the radius of the particle. For frequencies below 1000 GHz, the absorption efficiency is at least one order of magnitude smaller than the scattering efficiency. Therefore the extinction efficiency and the scattering efficiency are almost identical in this frequency range. The major maxima and minima of the scattering efficiency are called the *interference structure* and the irregular structure is called the *ripple structure*. The origin of the term *interference structure* lies in the interpretation of extinction as the interference between the incident and the forward-scattered light. The scattering efficiency Q_{sca} increases rapidly until the size parameter reaches approximately two for non-absorbing ice-particles. This means that for larger particles the maximum shifts towards lower frequencies. In later chapters, simulations for satellite limb measurements at 318 GHz will be shown. Figure 3.6 shows, that the scattering signal at this frequency should depend very much on the particle size. Another result of the Lorentz-Mie theory is that for a non-absorbing medium, for which the imaginary part of the refractive index equals zero, the scattering efficiency approaches an asymptotic value of two for large size parameters. This implies that the particle removes exactly twice the amount of energy that it can intercept. It includes the diffracted component, which passes by the particle, and additionally the radiation scattered by reflection and refraction inside the particle.

Figure 3.7 shows the extinction efficiency Q_{ext}, the scattering efficiency Q_{sca} and the absorption efficiency Q_{abs} of spherical water droplets of typical sizes. In contrast to ice particles, liquid water droplets mainly absorb microwave radiation. Below 1000 GHz scattering is negligibly small.

For polarized radiative transfer calculations, phase function, extinction coefficient, scattering coefficient and absorption coefficient are not sufficient. The VRTE Equation (1.42) shows that we need the phase matrix $\langle \boldsymbol{Z} \rangle$, the extinction matrix $\langle \boldsymbol{K} \rangle$ and the absorption coefficient vector $\langle \boldsymbol{a} \rangle$. The phase matrix represented in the scattering frame is commonly called scattering matrix $\boldsymbol{F}$. It follows from the Mie theory that the scattering matrix has only four independent matrix elements; it reduces to the simple form:

$$\boldsymbol{F}(\Theta) = \begin{pmatrix} F_{11}(\Theta) & F_{12}(\Theta) & 0 & 0 \\ F_{12}(\Theta) & F_{11}(\Theta) & 0 & 0 \\ 0 & 0 & F_{33}(\Theta) & F_{34}(\Theta) \\ 0 & 0 & -F_{34}(\Theta) & F_{33}(\Theta) \end{pmatrix}. \tag{3.7}$$

For spherical particles $\boldsymbol{F}$depends only on the scattering angle Θ, which is obvious for symmetry reasons.

3.3.3 T-matrix method

The deficit of the Lorentz-Mie theory is, that it provides scattering properties only for spherical particles. As shown in Section 3.1, ice particles are usually not spherical. The T-matrix method, which was initially introduced by Waterman (1965), can be applied for the computation of electromagnetic scattering by single, homogeneous, arbitrarily shaped particles. This original method is also known as the extended boundary condition method (EBCM). At present, the T-matrix approach is one of the most powerful and widely used tool for the computation of scattering by aspherical single and compounded

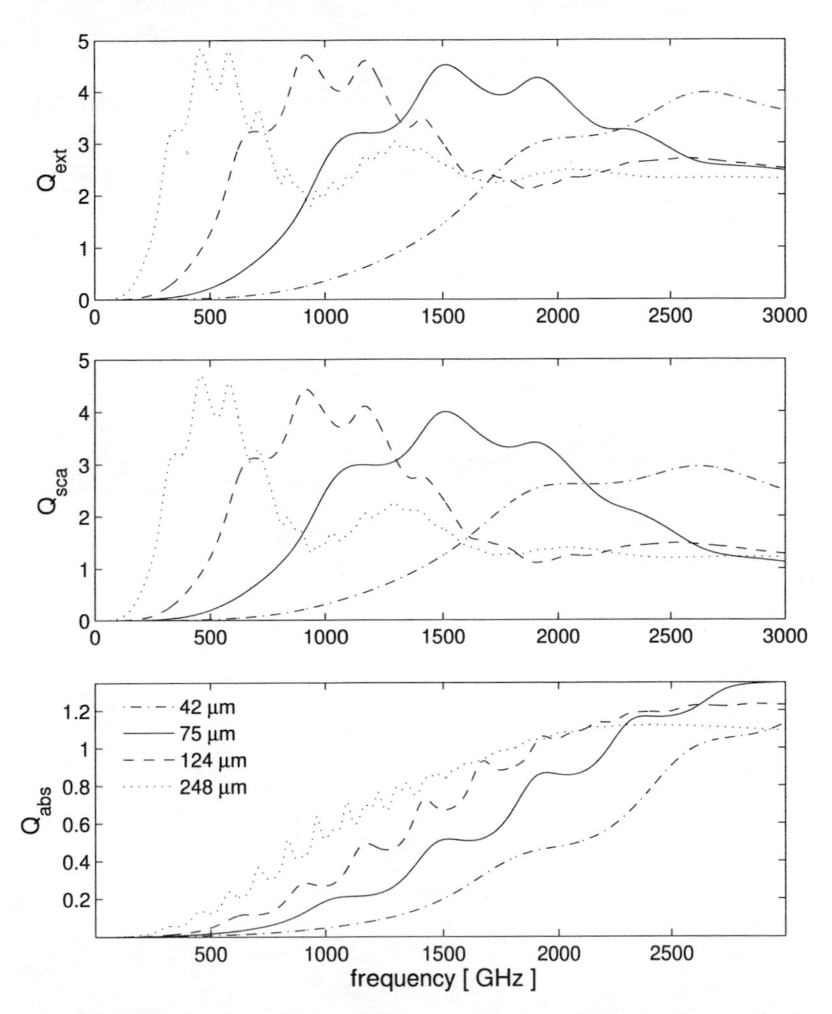

Figure 3.6: Extinction efficiency Q_{ext}, scattering efficiency Q_{sca} and absorption efficiency Q_{abs} in the microwave-wavelength region for spherical ice particles with radii of 42 µm, 75 µm, 124 µm and 248 µm.

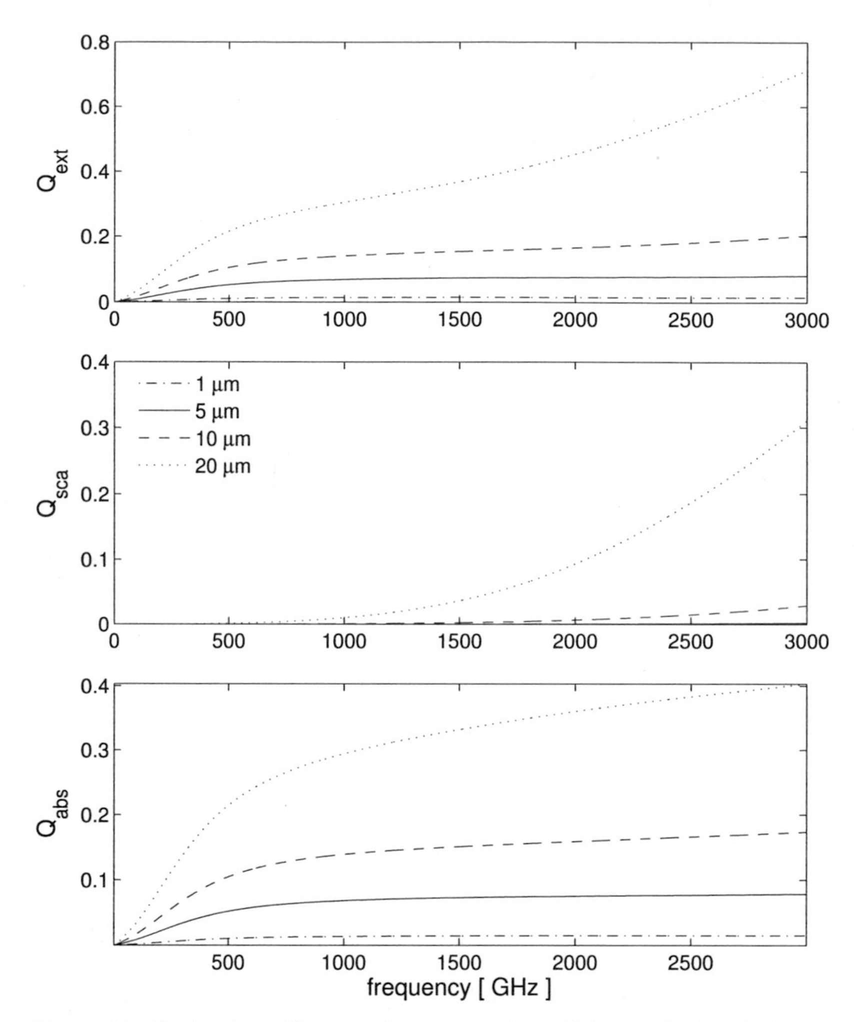

Figure 3.7: Extinction efficiency Q_{ext}, scattering efficiency Q_{sca} and absorption efficiency Q_{abs} in the microwave-wavelength region for spherical liquid water droplets with radii of $1\,\mu\text{m}$, $5\,\mu\text{m}$, $10\,\mu\text{m}$ and $20\,\mu\text{m}$.

particles. A detailed theoretical explanation can be found in the book by Mishchenko et al. (2002), who has developed several public domain programs which are available at http://www.giss.nasa.gov/~crmim. The T-matrix method needs the refractive index of ice as an input. Here the data from Warren (1984) was used to obtain the refractive index for the required frequencies and temperatures.

T-matrix program for randomly oriented, homogeneous and rotationally symmetric particles

The program for randomly oriented rotationally symmetric particles can be used for mono-disperse particles or for several analytical size distributions, e.g., the gamma distribution.

The total scattering matrix for the particle distribution is

$$\boldsymbol{F}(\Theta) = \begin{pmatrix} F_{11}(\Theta) & F_{12}(\Theta) & 0 & 0 \\ F_{12}(\Theta) & F_{22}(\Theta) & 0 & 0 \\ 0 & 0 & F_{33}(\Theta) & F_{34}(\Theta) \\ 0 & 0 & -F_{34}(\Theta) & F_{44}(\Theta) \end{pmatrix} = N \left\langle \boldsymbol{F}(\Theta) \right\rangle , \tag{3.8}$$

where N is the number of particles in a unit volume and $\langle \boldsymbol{F}(\Theta) \rangle$ is the ensemble-averaged scattering matrix per particle. In contrast to the scattering matrix for spherical particles, the matrix elements F_{11} and F_{22} as well as F_{33} and F_{44} are different, therefore we have now six instead of four independent elements. Like for spherical particles, the scattering matrix depends only on the scattering angle Θ.

The T-matrix program allows the computation of the scattering properties for non-spherical, rotationally symmetric particles. The shape of an aspherical particle is defined by its aspect ratio, which is the diameter of the particle divided by its length. Thus, a particle with an aspect ratio larger than one is a oblate particle and a particle with an aspect ratio smaller than one is a prolate particle.

Single scattering properties for mono-disperse spheroidal particles with an equal volume sphere radius of 75 µm are calculated. The results of the calculations for 318 GHz are shown in Figure 3.8. All the six phase matrix elements are presented. They are normalized by multiplication with $\frac{4\pi}{\langle C_{\mathrm{sca}} \rangle}$, where $\langle C_{\mathrm{sca}} \rangle$ is the ensemble averaged scattering cross-section. The grey lines correspond to the Mie calculation and the black lines correspond to the T-matrix calculations. The solid lines correspond to the normalized scattering matrix for spherical particles, which are identical to randomly oriented spheroids with an aspect ratio of 1.0. The dashed lines correspond to prolate spheroids with aspect ratios of 0.3 (thick) and 0.5 (thin). The dotted lines are the results for oblate spheroids with aspect ratios of 5.0 (thick) and 2.0 (thin). The Mie result corresponds to the T-matrix result for spherical particles, which means that the two methods are consistent. The matrix element F11 which corresponds to the phase function, shows that randomly oriented aspherical particles scatter slightly more radiation into the forward and slightly less radiation into the backward direction compared to spherical particles. The difference increases with increasing deformation. The matrix element $F21 = F12$ mainly determines the polarization state of the scattered radiation, since, for an unpolarized incident beam, the second Stokes component Q of the scattered beam corresponds to the product of $F21$ and the incoming intensity I^0. The plot shows, that maximal polarization occurs at a scattering angle of about 90ř and that Q is negative. The matrix element $F22$, which equals $F11$ for spherical particles, is for aspherical particles smaller than $F11$, especially in the backward direction. The absolute value of the element $F34 = F43$ is very small. $F33$ and $F44$ deviate only slightly from the result for spherical particles. Overall, the calculations for randomly oriented aspherical particles show, that at 318 GHz for a particle size of 75 µm the phase matrix is very similar to that for spherical particles. Therefore the dependence of simulated radiances on the particle shape at this frequency is expected to be much less than the dependence on the particle size.

T-matrix program for a particle in arbitrary orientation

All previous results were obtained for randomly oriented particles. In reality the ice crystals in cirrus clouds tend to be horizontally aligned. Another T-matrix program by Mishchenko (2000) is applicable for aspherical rotationally symmetric particles in arbitrary ori-

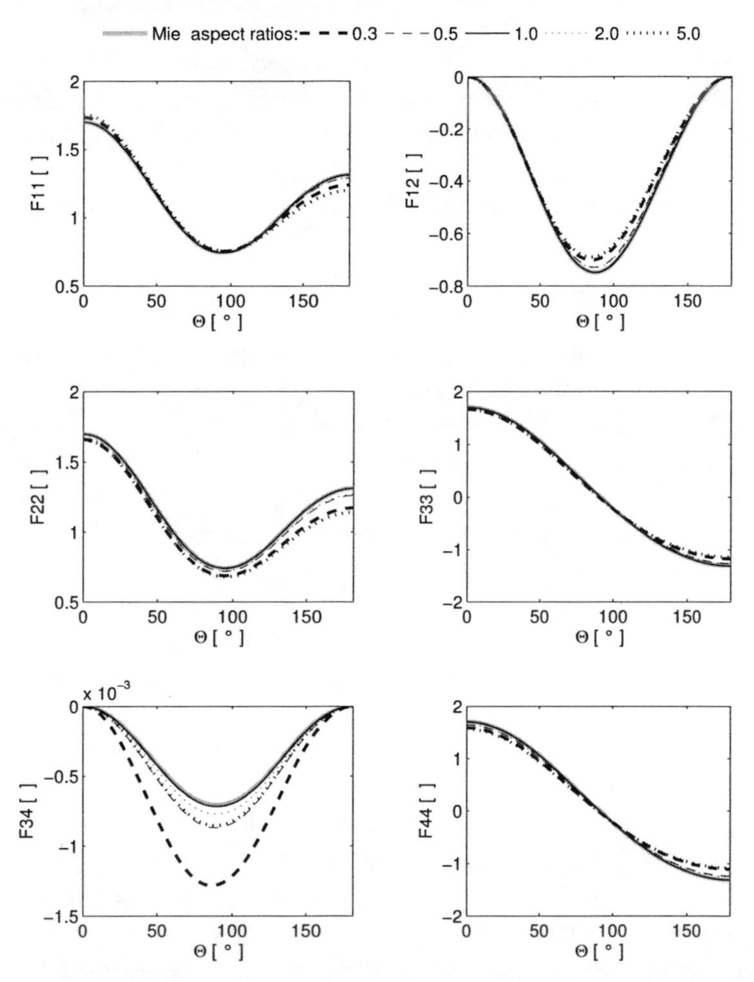

Figure 3.8: Grey line: Phase matrix elements calculated using Lorentz-Mie theory. Black lines: Orientation averaged phase matrix elements for spheroidal particles with different aspect ratios calculated using the T-matrix program for randomly oriented particles. The dashed lines correspond to prolate particles with aspect ratios of 0.3 (thick) and 0.5 (thin). The dotted lines correspond to oblate particles with aspect ratios of 2.0 (thin) and 5.0 (thick). The solid line corresponds to spherical particles. The equal volume sphere radius is 75 μm and the frequency is 318 GHz.

entation. This program has been used to calculate the phase matrix for cylindrical particles with different aspect ratios, which are horizontally oriented. The equal volume sphere radius is again 75 µm and the frequency of the calculation is 318 GHz. To be able to compare the results with Mie calculations, we have calculated the phase matrix for $\theta^{\text{inc}} = 0, \phi^{\text{inc}} = 0$, and $\phi^{\text{sca}} = 0$, so that it corresponds to the scattering matrix $\boldsymbol{F}$,

$$\boldsymbol{F}(\theta^{\text{sca}}) = \boldsymbol{Z}(\theta^{\text{sca}}, \phi^{\text{sca}} = 0, \theta^{\text{inc}} = 0, \phi^{\text{inc}} = 0). \tag{3.9}$$

The z-axes of the laboratory frame, in which the phase matrix is calculated, is chosen to be parallel to the incident direction ($\theta^{\text{inc}} = 0, \phi^{\text{inc}} = 0$). Thus θ^{sca} corresponds to the scattering angle Θ. The particle frame is defined in such a way that the z'-axes is parallel to the symmetry axes of the particle. Figure 3.9 shows that for a horizontally oriented plate the laboratory frame corresponds to the particle frame. For the horizontally oriented cylinder the z'-axes of the particle frame is rotated about the x-axes by 90° w.r.t. the z-axes. The orientation of the particles can be specified by appropriate Euler angles of rotation. The results of the calculations are presented in Figure 3.10. A cylinder with an aspect ratio of 1.0 gives almost identical results to the Mie calculation, when the symmetry axes corresponds to the z-axes. The results for the plates (dotted lines), which are oriented horizontally, are still similar. But the results for oriented cylinders (dashed lines) are very different. This is reasonable considering the symmetry. Like a sphere the plate has a circular geometrical cross-section when it is seen from the top. The cylinder, which is oriented horizontally, has a rectangular geometrical cross-section when seen from the top. The linear polarization is the difference between the vertical and the horizontal component of the intensity. As the horizontal and the vertical component of the electric field vector are perpendicular to the direction of propagation, the plate must have the same influence on both components for forward and backward scattering, because it is symmetric about the propagation direction of the radiation. Therefore $F21$, which is related to linear polarization, must be zero for $\Theta = 0°$ and for $\Theta = 180°$. As the cylindrical particle is not symmetric about the propagation direction, the scattered radiation is polarized, even

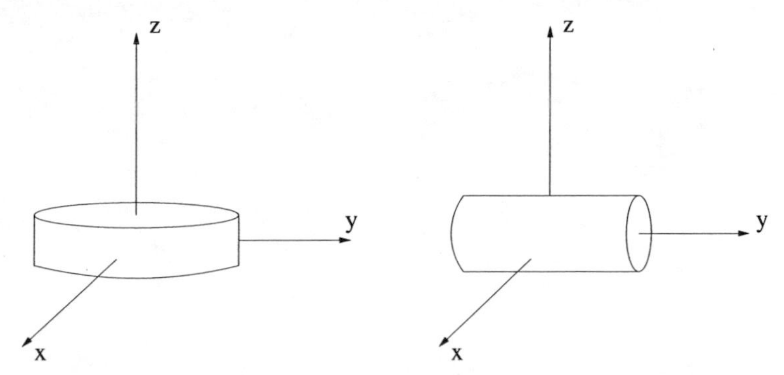

Figure 3.9: The left figure shows a horizontally oriented plate and the right figure shows a horizontally oriented cylinder.

for forward and backward scattering directions. For oriented particles the phase matrix does not depend only on the scattering angle but on the incident and scattered directions with respect to the particle orientation. For different directions, the phase matrix for the plates also deviates strongly from the Mie phase matrix.

This very short analysis of these special phase matrices shows that particle shape has a significant impact on the intensity and the polarization signal of microwave radiation in the atmosphere if the cirrus cloud particles are oriented.

3.3.4 Further methods

Other methods to calculate single scattering properties are the Discrete Dipole Approximation (DDA), the geometrical optics approximation, different extended T-matrix methods and several other methods, which are not discussed here.

Cloud particles are not so large that their optical properties in the microwave region could be calculated in the geometrical optics approximation. Therefore we have not considered this method.

The public domain DDA program DDSCAT developed by Draine and Flateau (2003) is available at http://www.astro.princeton.edu/~draine/DDSCAT.html. This program is widely used and tested for different types of particles. DDA can deal with any shapes of particles.

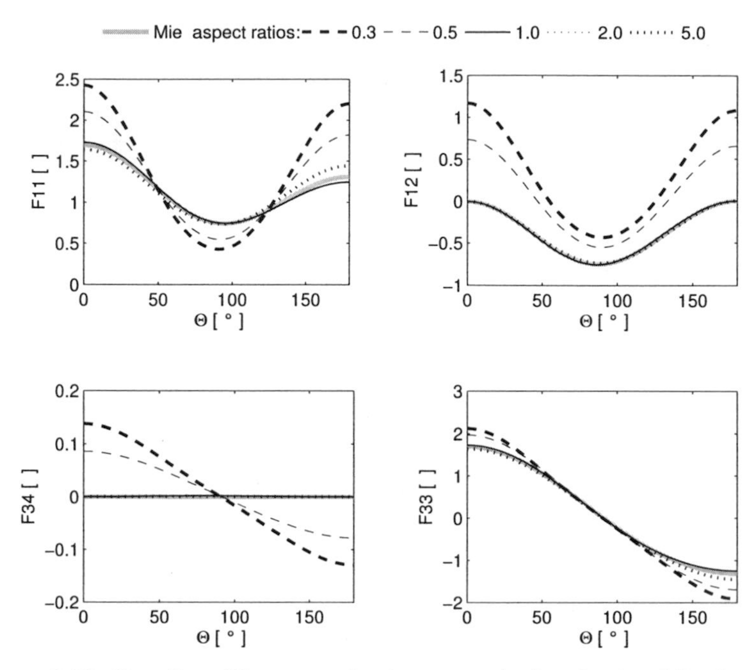

Figure 3.10: Grey line: Phase matrix elements calculated using Mie-theory. Black lines: Phase matrix elements for horizontally oriented plates/cylinders with different aspect ratios calculated using the T-matrix program for arbitrary oriented particles. The dashed lines correspond to cylinders with aspect ratios of 0.3 (thick) and 0.5 (thin). The dotted lines correspond to plates with aspect ratios of 2.0 (thin) and 5.0 (thick). The solid line corresponds to a cylinder with an aspect ratio of 1.0. The equal volume sphere radius is 75 μm and the frequency is 318 GHz.

The most noticeable shortcoming of DDA is its tremendous demand in computing time and memory. Compared to DDA, the T-matrix method is much faster and has much less demand in computer memory. However the Mishchenko code is designed only for particles with rotational symmetry and therefore it can be used only for a limited number of particle types. So far we have not used DDA for ARTS simulations, but it would be interesting to study the effect of particle shape for particles which are not rotationally symmetric and/or have extreme aspect ratios. This is planned in future studies.

The T-matrix results for spherical particles were compared to results obtained using the extended T-matrix code for aggregates, which was developed by Havemann and Baran (2001). The result is shown in Figure 3.11. The black lines show scattering, absorption and extinction efficiencies for spherical particles. The circles and cubes show the results for aggregates with aspect ratios of 1.0 and 4.0 respectively. For frequencies below 400 GHz there are only very small differences between the results of the two different aggregate particle types and the results for spherical particles. Only for frequencies about 500 GHz the difference becomes more significant, especially for the aggregate with an aspect ratio of 4.0. A drawback of the extended T-matrix program is that it provides only the cross sections, not the full extinction and phase matrices, which are required for polarized radiative transfer simulations. For this reason and because of the fact that the cross sections do not deviate much from the cross sections for spherical particles, it was decided not to use the extended T-matrix code as an input for ARTS.

3.4 Single scattering properties in the ARTS model

As shown in Section 3.1, clouds consist of a variety of particle sizes and shapes. Furthermore, the cloud particles can be oriented, in most cases they are horizontally aligned. It is not possible to model the clouds exactly as they are in nature, therefore we need some approximations. In ARTS different kinds of scattering media are implemented. One kind consists of randomly oriented particles, which allows very efficient computation of single scattering properties. Although this kind of scattering medium is only a special case, it provides a rather good numerical description of the scattering properties of clouds and is by far the most often used theoretical model in particle scattering theory. As the polarization signal of clouds depends significantly on the cloud particle orientation, there is also a kind of scattering media consisting of horizontally aligned particles.

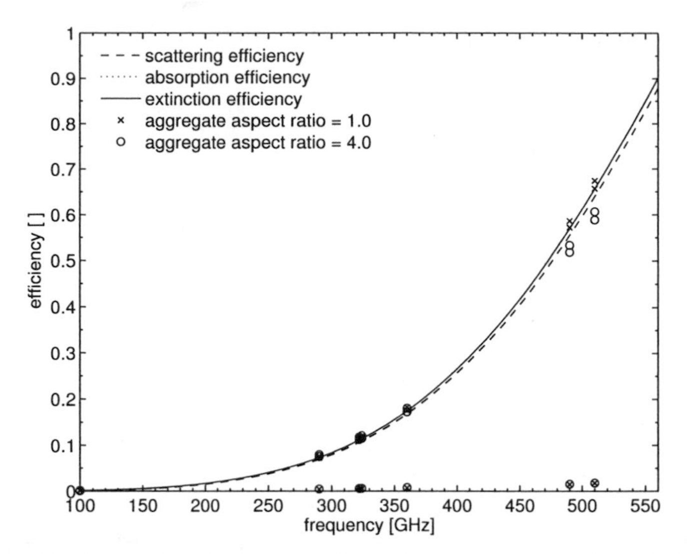

Figure 3.11: Comparison between different T-matrix codes. The black lines correspond to extinction (solid line), scattering (dashed line) and absorption (dotted line) efficiency obtained for spherical particles using the T-matrix code for randomly oriented particles. The crosses correspond to results obtained for aggregates with an aspect ratio of 1.0 and the circles are the results for aggregates with an aspect ratio of 4.0.

The single scattering properties are pre-calculated using the T-matrix method and stored in data-files. The structure of these data-files is shown in Table 3.1. The first field of the structure includes a value (*enum*), which characterizes the kind of scattering medium, i.e., whether the particles are arbitrarily oriented, randomly oriented or horizontally aligned. This information is needed, since different scattering media types are treated differently inside the program. The next field is a string which includes a short description of the scattering medium contained in the data-file. This should include information about how the single scattering properties were generated. After that, numerical grids for variables, on which the single scattering properties depend on, are stored. These are the frequency grid, the temperature grid, the zenith angle grid and the azimuth angle grid. Note that the same numerical grids are used for incident and scattered directions.

The last three fields contain the data. First the phase matrix data is stored as a seven-dimensional array, as the phase matrix depends on the frequency (ν), the temperature (T), the scattered direction (θ, ϕ), the incident direction (θ', ϕ') and it has in general sixteen matrix elements. The extinction matrix data and the absorption coefficient vector data are five-dimensional arrays. They have two dimensions less than the array holding the phase matrix data, since they are defined only for the forward direction.

Table 3.1: Structure of single scattering data files

Symbol	Type	Dimensions	Description
	enum		specification of particle type
	String		short description of particle type
ν	Vector	(ν)	frequency grid
T	Vector	(T)	temperature grid
θ	Vector	(θ)	zenith angle grid
ϕ	Vector	(ϕ)	azimuth angle grid
$\langle \boldsymbol{Z} \rangle$	7D Array	$(\nu, T, \theta, \phi, \theta', \phi', i)$	phase matrix
$\langle \boldsymbol{K} \rangle$	5D Array	$(\nu, T, \theta, \phi, i)$	extinction matrix
$\langle \boldsymbol{a} \rangle$	5D Array	$(\nu, T, \theta, \phi, i)$	absorption vector

The following three kinds of scattering media are implemented in ARTS so far. The number in brackets is the corresponding *enum*-value.

General case (p10): If the scattering medium does not have any symmetries, all sixteen elements of the phase matrix have to be stored. The individual phase matrices are calculated using Mishchenko's T-matrix code for single particles in fixed orientation, which is described in Section 3.3.3. The extinction matrix has in general seven independent elements and the absorption vector has four different elements.

Randomly oriented particles (p20): For this case the scattering medium is macroscopically isotropic and mirror-symmetric and we calculate the single scattering properties in the particle frame (Figure 3.3) using the T- matrix code for randomly oriented particles, which is described in Section 3.3.3. This reduces the size of the datafiles enormously, as the single scattering properties depend only on the scattering angle, not on four angles needed to describe incident and scattered directions. Furthermore the number of independent elements of the phase matrix, the extinction matrix and the absorption coefficient vector is reduced. Only six elements of the phase matrix are independent and the extinction matrix is diagonal, therefore only one element needs to be stored in the data files. Only the first element of the absorption vector is non-zero. Moreover, extinction and absorption are independent of the propagation direction. The only drawback is that the single scattering data has to be transformed from the particle frame representation to the laboratory frame representation. These transformations are described in detail in Appendix B.2.

Horizontally aligned plates and columns (p30): For particles that are azimuthally randomly oriented, one angular dimension of the phase matrix data array is redundant, as the phase matrix is independent of the incident azimuth angle. Furthermore, regarding the symmetry of this case, it can be shown that for the scattered directions only half of the angular grids are required. As for the general case, the fixed orientation T-matrix code for single scattering particles is used. The averaging over azimuthal orientations is done using the exact T-matrix averaging method of Mishchenko et al. (2000) for the extinction matrix, and by numerical integration for the phase matrix. The data is stored in the laboratory frame omitting the redundant data. Therefore this data does not need to be transformed. In order to use it for the radiative transfer equation, we only need appropriate reading and interpolation routines for this data format.

It is very convenient to use the PYTHON module PyARTS, which has been developed especially for ARTS and which is freely available at http://www.sat.uni-bremen.de/cgi-bin/cvsweb.cgi/PyARTS/. This module can be used to generate single scattering properties for horizontally aligned as well as for randomly oriented particles in the ARTS data-file-format. PyARTS has been developed by C. Davis, who has implemented the Monte Carlo scattering algorithm in ARTS (see Section 5.3).

3.5 Representation of the particle size distribution

The particle size has an important impact on the scattering and absorption properties of cloud particles as shown in Figure 3.6. Clouds contain a whole range of different particle sizes, which can be described by a size distribution giving the number of particles per unit volume per unit radius interval as a function of radius. It is most convenient to parameterize the size distribution by analytical functions, because in this case optical properties can be calculated much faster than for arbitrary size distributions. The T-matrix code for randomly oriented particles includes several types of analytical size distributions, e.g., the gamma distribution or the log-normal distribution. This section presents the size distribution parameterizations, which were used for the ARTS simulations included in this thesis.

3.5.1 Mono-disperse particle distribution

The most simple assumption is, that all particles in the cloud have the same size. In order to study scattering effects like polarization or the influence of particle shape, it makes sense to use this most simple assumption, because one can exclude effects coming from the particle size distribution itself. This simple assumption was made in the simulations presented in Chapter 7.

Along with the single scattering properties we need the particle number density field, which specifies the number of particles per cubic meter at each grid point, for ARTS scattering simulations. For a given IMC and mono-disperse particles the particle number density n^p is simply

$$n^p(\mathrm{IMC}, r) = \frac{\mathrm{IMC}}{m} = \frac{\mathrm{IMC}}{V\rho} = \frac{3}{4\pi}\frac{\mathrm{IMC}}{\rho r^3}, \tag{3.10}$$

where m is the mass of a particle, r is its equal volume sphere radius, ρ is its density, and V is its volume.

3.5.2 Gamma size distribution

A commonly used distribution for radiative transfer modeling in cirrus clouds is the *gamma distribution*

$$n(r) = ar^{\alpha} \exp(-br). \tag{3.11}$$

The dimensionless parameter α describes the width of the distribution. The other two parameters can be linked to the effective radius R_{eff} and the ice mass content IMC as follows:

$$b = \frac{\alpha + 3}{R_{\text{eff}}}, \tag{3.12}$$

$$a = \frac{\text{IMC}}{4/3\pi\rho b^{-(\alpha+4)}\Gamma[\alpha + 4]}, \tag{3.13}$$

where ρ is the density of the scattering medium and Γ is the gamma function. For cirrus clouds ρ corresponds to the bulk density of ice, which is 917 kg/m^3.

Generally, the effective radius R_{eff} is defined as the average radius weighted by the particle cross-section

$$R_{\text{eff}} = \frac{1}{\langle A \rangle} \int_{r_{\min}}^{r_{\max}} A(r) r n(r) \mathrm{d}r, \tag{3.14}$$

where A is the area of the geometric projection of a particle. The minimal and maximal particle sizes in the distribution are given by $r_{\min}$ and $r_{\max}$ respectively. In the case of spherical particles $A = \pi r^2$. The average area of the geometric projection per particle $\langle A \rangle$ is given by

$$\langle A \rangle = \frac{\int_{r_{\min}}^{r_{\max}} A(r) n(r) \mathrm{d}r}{\int_{r_{\min}}^{r_{\max}} n(r) \mathrm{d}r}. \tag{3.15}$$

The question is how well a gamma distribution can represent the true particle size distribution in radiative transfer calculations. This question is investigated by Evans et al. (1998). The authors come to the conclusion that a gamma distribution represents the distribution of realistic clouds quite well, provided that the parameters R_{eff}, IMC and α are chosen correctly. They show that setting $\alpha = 1$ and calculating only R_{eff} gave an agreement within 15% in 90% of the considered measurements obtained during the First ISCCP Regional Experiment (FIRE). Therefore, for all calculations including gamma size distributions for ice particles, $\alpha = 1$ was assumed. The results of these calculations are presented in Chapters 5 and 6.

The particle number density for size distributions is obtained by integration of the distribution function over all sizes:

$$n^p(\text{IMC}, R_{\text{eff}}) = \int_0^\infty n(r)\mathrm{d}r \tag{3.16}$$

$$= \int_0^\infty ar^\alpha \exp(-br)\mathrm{d}r = a\frac{\Gamma(\alpha+1)}{b^{\alpha+1}}. \tag{3.17}$$

After setting $\alpha = 1$, inserting Equation (3.13) and some simple algebra we obtain

$$n^p(\text{IMC}, R_{\text{eff}}) = \frac{2}{\pi}\frac{\text{IMC}}{\rho R_{\text{eff}}^3}. \tag{3.18}$$

Comparing Equations (3.10) and (3.18), we see that the particle number density for mono-disperse particles with a particle size of R is smaller than the particle number density for gamma distributed particles with $R_{\text{eff}} = R$. The reason is that in the gamma distribution most particles are smaller than R_{eff}.

3.5.3 Ice particle size parameterization by McFarquhar and Heymsfield

A more realistic parameterization of tropical cirrus ice crystal size distributions was derived by McFarquhar and Heymsfield (1997), who derived the size distribution as a function of temperature and IMC. The parameterization was made based on observations during the Central Equatorial Pacific Experiment (CEPEX). Smaller ice crystals with an equal volume sphere radius of less than 50 µm are parametrized as a sum of first-order gamma functions:

$$n(r) = \frac{12 \cdot \text{IMC}_{<50}\alpha_{<50}^5 r}{\pi\rho\Gamma(5)} \exp(-2\alpha_{<50} r), \tag{3.19}$$

where $\alpha_{<50}$ is a parameter of the distribution, and $\text{IMC}_{<50}$ is the mass of all crystals smaller than 50 µm in the observed size distribution.

Large ice crystals are represented better by a log-normal function

$$n(r) = \frac{3 \cdot \mathrm{IMC}_{>50}}{\pi^{3/2}\rho\sqrt{2}\exp(3\mu_{>50} + (9/2)\sigma^2_{>50}) r \sigma_{>50} r_0^3} \cdot \exp\left[-\frac{1}{2}\left(\frac{\log\frac{2r}{r_0} - \mu_{>50}}{\sigma_{>50}}\right)^2\right], \tag{3.20}$$

where $\mathrm{IMC}_{>50}$ is the mass of all ice crystals greater than 50 µm in the observed size distribution, $r_0 = 1\,\mu\mathrm{m}$ is a parameter used to ensure that the equation does not depend on the choice of unit for r, $\sigma_{>50}$ is the geometric standard deviation of the distribution, and $\mu_{>50}$ is the location of the mode of the log-normal distribution. The fitted parameters of the distribution can be looked up in the article by McFarquhar and Heymsfield (1997). The particle number density field is obtained by numerical integration over a discrete set of size bins. This parameterization of particle size has been implemented in the PyARTS package, which was introduced in Section 3.4. Using PyARTS one can calculate the size distributions, the corresponding single scattering properties and the particle number density fields for given IMC and temperature. Calculations using this parameterization of cloud particle sizes are presented in Chapter 8.

4 The DOIT scattering model

The Discrete Ordinate ITerative (DOIT) method is one of the scattering algorithms in ARTS. Besides the DOIT method a backward Monte Carlo scheme has been implemented (see Section 5.3). The DOIT method is unique because a discrete ordinate iterative method is used to solve the scattering problem in a 3D spherical atmosphere. A literature review about scattering models for the microwave region, which is presented in Appendix A, shows that former implementations of discrete ordinate schemes are only applicable for (1D-)plane-parallel or 3D-cartesian atmospheres. All of these algorithms can not be used for the simulation of limb radiances. A description of the DOIT method, similar to what is presented in this chapter, has been published in Emde et al. (2004a).

4.1 The discrete ordinate iterative method

4.1.1 Radiation field

The Stokes vector depends on the position in the cloud box and on the propagation direction specified by the zenith angle (θ) and the azimuth angle (ϕ). All these dimensions are discretized inside the model; five numerical grids are required to represent the radiation field $\mathcal{I}$:

$$
\begin{aligned}
\vec{p} &= \{p_1, p_2, ..., p_{N_p}\}, \\
\vec{\alpha} &= \{\alpha_1, \alpha_2, ..., \alpha_{N_\alpha}\}, \\
\vec{\beta} &= \{\beta_1, \beta_2, ..., \beta_{N_\beta}\}, \\
\vec{\theta} &= \{\theta_1, \theta_2, ..., \theta_{N_\theta}\}, \\
\vec{\phi} &= \{\phi_1, \phi_2, ..., \phi_{N_\phi}\}.
\end{aligned}
\tag{4.1}
$$

Here $\vec{p}$ is the pressure grid, $\vec{\alpha}$ is the latitude grid and $\vec{\beta}$ is the longitude grid. The radiation field is a set of Stokes vectors ($N_p \times N_\alpha \times N_\beta \times N_\theta \times N_\phi$ elements) for all combinations of positions and directions:

$$\begin{aligned}\mathcal{I} = \{&\boldsymbol{I}_1(p_1, \alpha_1, \beta_1, \theta_1, \phi_1), \boldsymbol{I}_2(p_2, \alpha_1, \beta_1, \theta_1, \phi_1), ..., \\ &\boldsymbol{I}_{N_p \times N_\alpha \times N_\beta \times N_\theta \times N_\phi}(p_{N_p}, \alpha_{N_\alpha}, \beta_{N_\beta}, \theta_{N_\theta}, \phi_{N_\phi})\}.\end{aligned} \tag{4.2}$$

In the following we will use the notation

$$\mathcal{I} = \{\boldsymbol{I}_{ijklm}\} \qquad \begin{aligned} i &= 1 \dots N_p \\ j &= 1 \dots N_\alpha \\ k &= 1 \dots N_\beta. \\ l &= 1 \dots N_\theta \\ m &= 1 \dots N_\phi \end{aligned} \tag{4.3}$$

4.1.2 Vector radiative transfer equation solution

Figure 4.1 shows a schematic of the iterative method, which is applied to solve the vector radiative transfer equation (1.42).

The *first guess field*

$$\mathcal{I}^{(0)} = \left\{ \boldsymbol{I}^{(0)}_{ijklm} \right\}, \tag{4.4}$$

is partly determined by the boundary condition given by the radiation coming from the clear sky part of the atmosphere traveling into the cloud box. Inside the cloud box an arbitrary field can be chosen as a first guess. In order to minimize the number of iterations it should be as close as possible to the solution field.

The next step is to solve the scattering integrals

$$\left\langle \boldsymbol{S}^{(0)}_{ijklm} \right\rangle = \int_{4\pi} \mathrm{d}\boldsymbol{n}' \left\langle \boldsymbol{Z}_{ijklm} \right\rangle \boldsymbol{I}^{(0)}_{ijklm}, \tag{4.5}$$

using the first guess field, which is now stored in a variable reserved for the *old radiation field*. For the integration we use equidistant angular grids in order to save computation time (cf. Section 4.3). The radiation field, which is generally defined on finer angular grids ($\vec{\phi}, \vec{\theta}$),

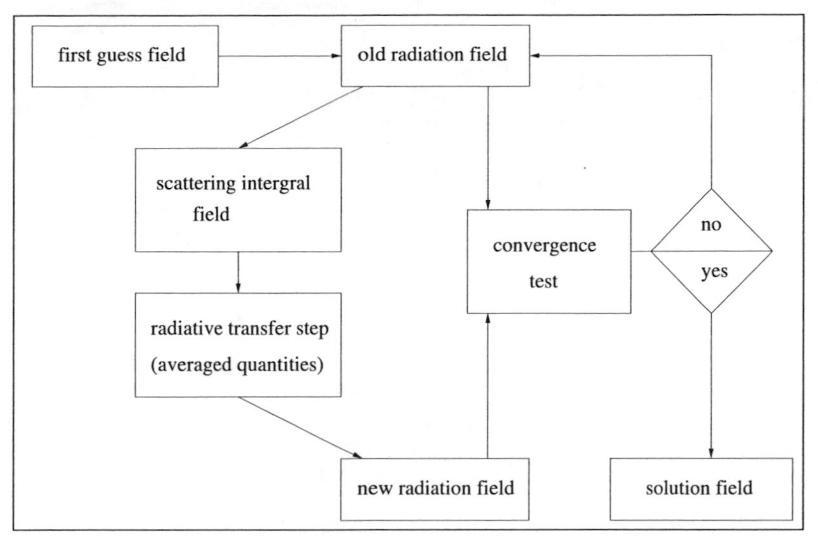

Figure 4.1: Schematic of the iterative method to solve the VRTE in the cloud box.

is interpolated on the equidistant angular grids. The integration is performed over all incident directions $\boldsymbol{n}'$ for each propagation direction $\boldsymbol{n}$. The evaluation of the scattering integral is done for all grid points inside the cloud box. The obtained integrals are interpolated on $\vec{\phi}$ and $\vec{\theta}$. The result is the first guess *scattering integral field* $\mathcal{S}^0$:

$$\mathcal{S}^{(0)} = \left\{ \left\langle \boldsymbol{S}^{(0)}_{ijklm} \right\rangle \right\}. \tag{4.6}$$

Figure 4.2 shows a propagation path step from a grid point $\boldsymbol{P} = (p_i, \alpha_j, \beta_k)$ into direction $\boldsymbol{n} = (\theta_l, \phi_m)$. The radiation arriving at $\boldsymbol{P}$ from the direction $\boldsymbol{n}'$ is obtained by solving the linear differential equation:

$$\frac{\mathrm{d}\boldsymbol{I}^{(1)}}{\mathrm{d}s} = -\overline{\langle \boldsymbol{K} \rangle} \boldsymbol{I}^{(1)} + \overline{\langle \boldsymbol{a} \rangle}\, \bar{B} + \overline{\left\langle \boldsymbol{S}^{(0)} \right\rangle}, \tag{4.7}$$

where $\overline{\langle \boldsymbol{K} \rangle}$, $\overline{\langle \boldsymbol{a} \rangle}$, $\bar{B}$ and $\overline{\left\langle \boldsymbol{S}^{(0)} \right\rangle}$ are *averaged quantities.* This equation can be solved analytically for constant coefficients. Multi-linear interpolation gives the quantities $\boldsymbol{K}', \boldsymbol{a}', \boldsymbol{S}'$ and T' at the intersection point $\boldsymbol{P}'$. To calculate the radiative transfer from $\boldsymbol{P}'$ towards $\boldsymbol{P}$ all

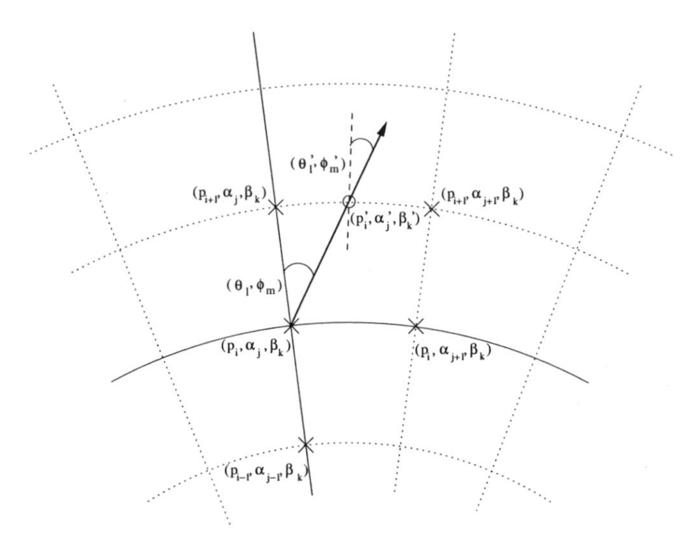

Figure 4.2: Path from a grid point $((p_i, \alpha_j, \beta_k)$ - $(\times))$ to the intersection point $((p_i', \alpha_j', \beta_k')$ - $(\circ))$ with the next grid cell boundary. Viewing direction is specified by (θ_l, ϕ_m) at $(\times)$ or (θ_l', ϕ_m') at $(\circ)$.

quantities are approximated by taking the averages between the values at $\boldsymbol{P}'$ and $\boldsymbol{P}$. The average value of the temperature is used to get the averaged Planck function $\bar{B}$.

The solution of Equation (4.7) is found analytically using a matrix exponential approach (see Appendix B.1):

$$\boldsymbol{I}^{(1)} = e^{-\overline{\langle \boldsymbol{K} \rangle} s} \boldsymbol{I}^{(0)} \left(\mathbb{I} - e^{-\overline{\langle \boldsymbol{K} \rangle} s} \right) \overline{\langle \boldsymbol{K} \rangle}^{-1} \left(\overline{\langle \boldsymbol{a} \rangle}\, \bar{B} + \overline{\left\langle \boldsymbol{S}^{(0)} \right\rangle} \right), \quad (4.8)$$

where $\mathbb{I}$ denotes the identity matrix and $\boldsymbol{I}^{(0)}$ the initial Stokes vector. The *radiative transfer step* from $\boldsymbol{P}'$ to $\boldsymbol{P}$ is calculated, therefore $\boldsymbol{I}^{(0)}$ is the incoming radiation at $\boldsymbol{P}'$ into direction (θ_l', ϕ_m'), which is the first guess field interpolated on $\boldsymbol{P}'$. This radiative transfer step calculation is done for all points inside the cloud box in all directions. The resulting set of Stokes vectors ($\boldsymbol{I}^{(1)}$ for all points in all directions) is the first order iteration field $\mathcal{I}^{(1)}$:

$$\mathcal{I}^{(1)} = \left\{ \boldsymbol{I}^{(1)}_{ijklm} \right\}. \quad (4.9)$$

The first order iteration field is stored in a variable reserved for the *new radiation field*.

In the *convergence test* the *new radiation field* is compared to the *old radiation field*. For the difference field, the absolute values of all Stokes vector elements for all cloud box positions are calculated. If one of the differences is larger than a requested accuracy limit, the convergence test is not fulfilled. The user can define different convergence limits for the different Stokes components.

If the convergence test is not fulfilled, the first order iteration field is copied to the variable holding the *old radiation field*, and is then used to evaluate again the scattering integral at all cloud box points:

$$\left\langle \boldsymbol{S}^{(1)}_{ijklm} \right\rangle = \int_{4\pi} \mathrm{d}\boldsymbol{n}' \left\langle \boldsymbol{Z} \right\rangle \boldsymbol{I}^{(1)}_{ijklm}. \quad (4.10)$$

The second order iteration field

$$\mathcal{I}^{(2)} = \left\{ \boldsymbol{I}^{(2)}_{ijklm} \right\}, \quad (4.11)$$

is obtained by solving

$$\frac{\mathrm{d}\boldsymbol{I}^{(2)}}{\mathrm{d}s} = -\overline{\langle \boldsymbol{K} \rangle} \boldsymbol{I}^{(2)} + \overline{\langle \boldsymbol{a} \rangle}\, \bar{B} + \overline{\left\langle \boldsymbol{S}^{(1)} \right\rangle}, \quad (4.12)$$

for all cloud box points in all directions. This equation contains already the averaged values and is valid for specified positions and directions.

As long as the convergence test is not fulfilled the scattering integral fields and higher order iteration fields are calculated alternately.

We can formulate a differential equation for the n-th order iteration field. The scattering integrals are given by

$$\left\langle \boldsymbol{S}^{(n-1)}_{ijklm} \right\rangle = \int_{4\pi} \mathrm{d}\boldsymbol{n}' \left\langle \boldsymbol{Z} \right\rangle \boldsymbol{I}^{(n-1)}_{ijklm}, \quad (4.13)$$

and the differential equation for a specified grid point into a specified direction is

$$\frac{\mathrm{d}\boldsymbol{I}^{(n)}}{\mathrm{d}s} = -\overline{\langle \boldsymbol{K} \rangle} \boldsymbol{I}^{(n)} + \overline{\langle \boldsymbol{a} \rangle}\, \bar{B} + \overline{\left\langle \boldsymbol{S}^{(n-1)} \right\rangle}. \quad (4.14)$$

Thus the *n-th order iteration field*

$$\mathcal{I}^{(n)} = \left\{ \boldsymbol{I}^{(n)}_{ijklm} \right\}, \quad (4.15)$$

is given by

$$\boldsymbol{I}^{(n)} = e^{-\overline{\langle \boldsymbol{K} \rangle} s} \cdot \boldsymbol{I}^{(n-1)} (\mathbb{I} - e^{-\overline{\langle \boldsymbol{K} \rangle} s}) \overline{\langle \boldsymbol{K} \rangle}^{-1} (\overline{\langle \boldsymbol{a} \rangle} \bar{B} + \overline{\left\langle \boldsymbol{S}^{(n-1)} \right\rangle}), \tag{4.16}$$

for all cloud box points and all directions defined in the numerical grids.

If the convergence test

$$\left| \boldsymbol{I}^{(N)}_{ijklm} (p_i, \alpha_j, \beta_k, \theta_l, \phi_m) - \boldsymbol{I}^{(N-1)}_{ijklm} (p_i, \alpha_j, \beta_k, \theta_l, \phi_m) \right| < \boldsymbol{\epsilon}, \tag{4.17}$$

is fulfilled, a solution to the vector radiative transfer equation (1.42) has been found:

$$\mathcal{I}^{(N)} = \left\{ \boldsymbol{I}^{(N)}_{ijklm} \right\}. \tag{4.18}$$

4.1.3 Scalar radiative transfer equation solution

In analogy to the *scattering integral* vector field the scalar scattering integral field is obtained:

$$\left\langle S^{(0)}_{ijklm} \right\rangle = \int_{4\pi} \mathrm{d}\boldsymbol{n}' \left\langle Z_{11} \right\rangle I^{(0)}_{ijklm}. \tag{4.19}$$

The *scalar radiative transfer* equation (1.51) with a fixed scattering integral is

$$\frac{\mathrm{d}I^{(1)}}{\mathrm{d}s} = - \left\langle K_{11} \right\rangle I^{(1)} + \left\langle a_1 \right\rangle B + \left\langle S^{(0)} \right\rangle. \tag{4.20}$$

Assuming constant coefficients this equation is solved analytically after averaging extinction coefficients, absorption coefficients, scattering vectors and the temperature. The averaging procedure is done analogously to the procedure described for solving the VRTE. The solution of the averaged differential equation is

$$I^{(1)} = I^{(0)} e^{-\overline{\langle K_{11} \rangle} s} + \frac{\overline{\langle a_1 \rangle} \bar{B} + \overline{\langle S^{(0)} \rangle}}{\overline{\langle K_{11} \rangle}} \left(1 - e^{-\overline{\langle K_{11} \rangle} s} \right), \tag{4.21}$$

where $I^{(0)}$ is obtained by interpolating the initial field, and $\overline{\langle K_{11} \rangle}$, $\overline{\langle a_1 \rangle}$, $\bar{B}$ and $\overline{\langle S^{(0)} \rangle}$ are the averaged values for the extinction coefficient, the absorption coefficient, the Planck function and the scattering integral respectively. Applying this equation leads to the first iteration scalar intensity field, consisting of the intensities $I^{(1)}$ at all points in the cloud box for all directions.

As the solution to the vector radiative transfer equation, the solution to the scalar radiative transfer equation is found numerically by the same iterative method. The convergence test for the scalar equation compares the values of the calculated intensities of two successive iteration fields.

4.1.4 Single scattering approximation

The DOIT method uses the single scattering approximation, which means that for one propagation path step the optical depth is assumed to be much less than one so that multiple-scattering can be neglected along this propagation path step. It is possible to choose a rather coarse grid inside the cloud box. The user can define a limit for the maximum propagation path step length. If a propagation path step from one grid cell to the intersection point with the next grid cell boundary is greater than this value, the path step is divided in several steps such that all steps are less than the maximum value. The user has to make sure that the optical depth due to cloud particles for one propagation path sub-step is is sufficiently small to assume single scattering. The maximum optical depth due to ice particles is

$$\tau_{max} = \langle \boldsymbol{K}^p \rangle \cdot \Delta s, \tag{4.22}$$

where Δs is the length of a propagation path step. In all simulations presented in this thesis $\tau_{max} \ll 0.01$ is assumed. This threshold value is also used in Czekala (1999a). The radiative transfer calculation is done along the propagation path through one grid cell. All coefficients of the VRTE are interpolated linearly on the propagation path points.

4.2 Sequential update

In the previous Section 4.1 the iterative solution method for the VRTE has been described. For each grid point inside the cloud box the intersection point with the next grid cell boundary is determined in each viewing direction. After that, all the quantities involved in the VRTE are interpolated onto this intersection point. As described in the sections above, the intensity field of the previous iteration is taken to obtain the Stokes vector at the intersection point. Suppose that there are N pressure levels inside the cloud box. If the radiation field is updated taking into account for each grid point only the adjacent grid cells, at least N-1 iterations are required until the scattering effect from the lower-most pressure level has propagated throughout the cloud box up to the uppermost pressure level. From these considerations, it follows, that the number of iterations depends on the number of grid points inside the cloud box. This means that the original method is very ineffective where a fine resolution inside the cloud box is required to resolve the cloud inhomogeneities.

A solution to this problem is the "sequential update of the radiation field", which is shown schematically in Figure 4.3. For simplicity it will be explained in detail for a 1D cloud box. We divide the update of the radiation field, i.e., the radiative transfer step calculations for all positions and directions inside the cloud box, into three parts: Update for "up-looking" zenith angles ($0° \leq \theta_{\mathrm{up}} \leq 90°$), for "down-looking" angles ($\theta_{\mathrm{limit}} \leq \theta_{\mathrm{down}} \leq 180°$) and for "limb-looking" angles ($90° < \theta_{\mathrm{limb}} < \theta_{\mathrm{limit}}$). The "limb-looking" case is needed, because for angles between 90° and θ_{limit} the intersection point is at the same pressure level as the observation point. The limiting angle θ_{limit} is calculated geometrically. Note that the propagation direction of the radiation is opposite to the viewing direction or the direction of the line of sight, which is indicated by the arrows. In the 1D case the radiation field is a set of Stokes vectors each of which depend upon the position and direction:

$$\mathcal{I} = \{\boldsymbol{I}(p_i, \theta_l)\} . \tag{4.23}$$

The *boundary condition* for the calculation is the incoming radiation

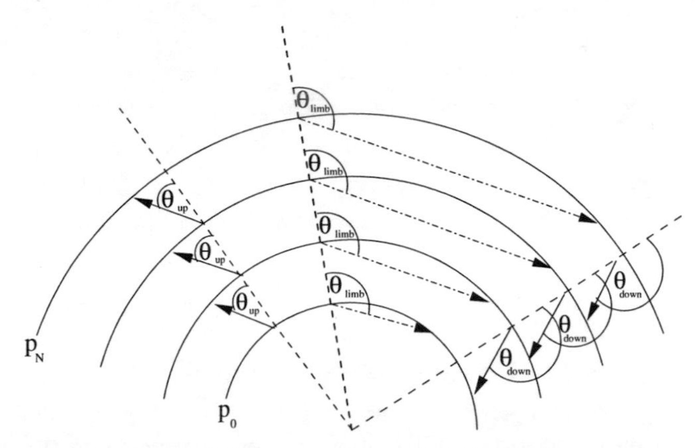

Figure 4.3: Schematic of the sequential update (1D) showing the three different parts: "up-looking" corresponds to zenith angles θ_{up}, "limb-looking" corresponds to θ_{limb} "down-looking" corresponds to θ_{down}.

field on the cloud box boundary $\mathcal{I}^{bd}$:

$$\mathcal{I}^{bd} = \{\boldsymbol{I}(p_i, \theta_l)\} \text{ where } p_i = p_N \,\forall\, \theta_l \in [0, \theta_{\mathrm{limit}}] \\ p_i = p_0 \,\forall\, \theta_l \in (\theta_{\mathrm{limit}}, 180°], \tag{4.24}$$

where p_0 and p_N are the pressure coordinates of the lower and upper cloud box boundaries respectively. For down-looking directions, the intensity field at the lower-most cloud box boundary and for up- and limb-looking directions the intensity field at the uppermost cloud box boundary are the required boundary conditions respectively.

4.2.1 Up-looking directions

The first step of the sequential update is to calculate the intensity field for the pressure coordinate p_{N-1}, the pressure level below the uppermost boundary, for all up-looking directions. Radiative transfer steps are calculated for paths starting at the uppermost boundary and propagating to the $(N-1)$ pressure level. The required input for this radiative transfer step are the averaged coefficients of the uppermost cloud box layer and the Stokes vectors at the uppermost boundary for all up-looking directions. These are obtained by interpolating the boundary condition $\mathcal{I}^{bd}$ on the appropriate zenith angles. Note that the zenith angle of the propagation path for the observing direction θ_l does not equal θ'_l at the intersection point due to the spherical geometry. If θ_l is close to 90° this difference is most significant.

To calculate the intensity field for the pressure coordinate p_{N-2}, we repeat the calculation above. We have to calculate a radiative transfer step from the $(N-1)$ to the $(N-2)$ pressure level. As input we need the interpolated intensity field at the $(N-1)$ pressure level, which has been calculated in the last step.

For each pressure level $(m-1)$ we take the interpolated field of the layer above $(\mathcal{I}(p_m)^{(1)})$. Using this method, the scattering influence from particles in the upper-most cloud box layer can propagate during one iteration down to the lower-most layer. This means that the number of iterations does not scale with the number of pressure levels, which would be the case without sequential update.

The radiation field at a specific point in the cloud box is obtained by solving Equation (4.8). For up-looking directions at position p_{m-1} we may write:

$$\begin{aligned}\boldsymbol{I}\left(p_{m-1},\theta_{\mathrm{up}}\right)^{(1)} &= e^{-\overline{\langle \boldsymbol{K}(\theta_{\mathrm{up}})\rangle} s}\boldsymbol{I}\left(p_m,\theta_{\mathrm{up}}\right)^{(1)} \\ &+ \left(\mathbb{I} - e^{-\overline{\langle \boldsymbol{K}(\theta_{\mathrm{up}})\rangle} s}\right)\overline{\langle \boldsymbol{K}(\theta_{\mathrm{up}})\rangle}^{-1}\left(\overline{\langle \boldsymbol{a}(\theta_{\mathrm{up}})\rangle}\,\bar{B} + \overline{\left\langle \boldsymbol{S}\left(\theta_{\mathrm{up}}\right)^{(0)}\right\rangle}\right).\end{aligned} \tag{4.25}$$

For simplification we write

$$\boldsymbol{I}(p_{m-1},\theta_{\mathrm{up}})^{(1)} = \boldsymbol{A}(\theta_{\mathrm{up}})\boldsymbol{I}\left(p_m,\theta_{\mathrm{up}}\right)^{(1)} + \boldsymbol{B}(\theta_{\mathrm{up}}). \tag{4.26}$$

Solving this equation sequentially, starting at the top of the cloud and finishing at the bottom, we get the updated radiation field for all up-looking angles.

$$\mathcal{I}(p_i,\theta_{\mathrm{up}})^{(1)} = \left\{\boldsymbol{I}^{(1)}\left(p_i,\theta_l\right)\right\} \qquad \forall\,\theta_l \in [0,90°]. \tag{4.27}$$

4.2.2 Down-looking directions

The same procedure is done for down-looking directions. The only difference is that the starting point is the lower-most pressure level

p_1 and the incoming clear sky field at the lower cloud box boundary, which is interpolated on the required zenith angles, is taken as boundary condition. The following equation is solved sequentially, starting at the bottom of the cloud box and finishing at the top:

$$\boldsymbol{I}(p_m, \theta_{\text{down}})^{(1)} = \boldsymbol{A}(\theta_{\text{down}})\boldsymbol{I}\left(p_{m-1}, \theta_{\text{down}}\right)^{(1)} + \boldsymbol{B}(\theta_{\text{down}}). \quad (4.28)$$

This yields the updated radiation field for all down-looking angles.

$$\mathcal{I}(p_i, \theta_{\text{down}})^{(1)} = \left\{\boldsymbol{I}^{(1)}\left(p_i, \theta_l\right)\right\} \qquad \forall\, \theta_l \in [\theta_{\text{limit}}, 180°]. \quad (4.29)$$

4.2.3 Limb directions

A special case for limb directions, which correspond to angles slightly above 90° had to be implemented. If the tangent point is part of the propagation path step, the intersection point is exactly at the same pressure level as the starting point. In this case the linearly interpolated clear sky field is taken as input for the radiative transfer calculation, because we do not have an already updated field for this pressure level:

$$\boldsymbol{I}(p_m, \theta_{\text{limb}})^{(1)} = \boldsymbol{A}(\theta_{\text{limb}})\boldsymbol{I}\left(p_m, \theta_{\text{limb}}\right)^{(0)} + \boldsymbol{B}(\theta_{\text{limb}}) \quad (4.30)$$

By solving this equation the missing part of the updated radiation field is obtained

$$\mathcal{I}(p_i, \theta_{\text{limb}})^{(1)} = \left\{\boldsymbol{I}\left(p_i, \theta_l\right)\right\} \qquad \forall\, \theta_l \in]90°, \theta_{\text{limit}}[\quad (4.31)$$

For all iterations the sequential update is applied. Using this method the number of iterations depends only on the optical thickness of the cloud or on the number of multiple-scattering events, not on the number of pressure levels.

How the sequential update is performed in the 3D model is described in Eriksson et al. (2004).

4.3 Grid optimization and interpolation methods

The accuracy of the DOIT method depends very much on the discretization of the zenith angle. The reason is that the intensity field strongly increases at about $\theta = 90°$. For angles below 90° ("up-looking" directions) the intensity is very small compared to angles above 90° ("down-looking" directions), because the thermal emission from the lower atmosphere and from the ground is much larger than thermal emission from trace gases in the upper atmosphere. Figure 4.4 shows an example intensity field as a function of zenith angle for different pressure levels inside a cloud box, which is placed from 7.3 to 12.7 km altitude, corresponding to pressure limits of 411 hPa and 188 hPa respectively. The cloud box includes 27 pressure levels. The frequency of the sample calculation was 318 GHz. A midlatitude-summer scenario including water vapor, ozone, nitrogen and oxygen was used. The atmospheric data was taken from the FASCOD (Anderson et al., 1986) and the spectroscopic data was obtained from the HITRAN database (Rothman et al., 1998). For simplicity this 1D set-up was chosen for all sample calculations in this section. As the intensity (or the Stokes vector) at the intersection point of a propagation path is obtained by interpolation, large interpolation errors can occur for zenith angles of about 90° if the zenith angle grid discretization is too coarse. Taking a very fine equidistant zenith angle grid leads to very long computation times. Therefore a zenith angle grid optimization method is required.

For the computation of the scattering integral it is possible to take a much coarser zenith angle resolution without losing accuracy. It does not make sense to use the zenith angle grid, which is optimized to represent the radiation field with a certain accuracy. The integrand is the product of the phase matrix and the radiation field. The peaks of the phase matrices can be at any zenith angle, depending on the incoming and the scattered directions. The multiplication smooths out both the radiation field increase at 90° and the peaks of the phase matrices. Test calculations have shown that an increment of 10° is sufficient. Taking the equidistant grid saves the computation time of the scattering integral to a very large extent, because much less grid points are required.

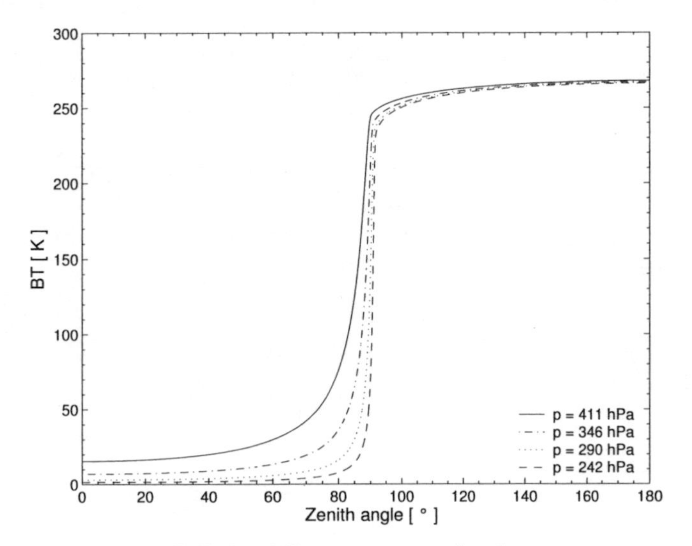

Figure 4.4: Intensity field for different pressure levels.

4.3.1 Zenith angle grid optimization

As a reference field for the grid optimization the DOIT method is applied for an empty cloud box using a very fine zenith angle grid. The grid optimization routine finds a reduced zenith angle grid which can represent the intensity field with the desired accuracy. It first takes the radiation at 0° and 180° and interpolates between these two points on all grid points contained in the fine zenith angle grid for all pressure levels. Then the differences between the reference radiation field and the interpolated field are calculated. The zenith angle grid point, where the difference is maximal is added to 0° and 180°. After that the radiation field is interpolated between these three points forming part of the reduced grid and again the grid point with the maximum difference is added. Using this method more and more grid points are

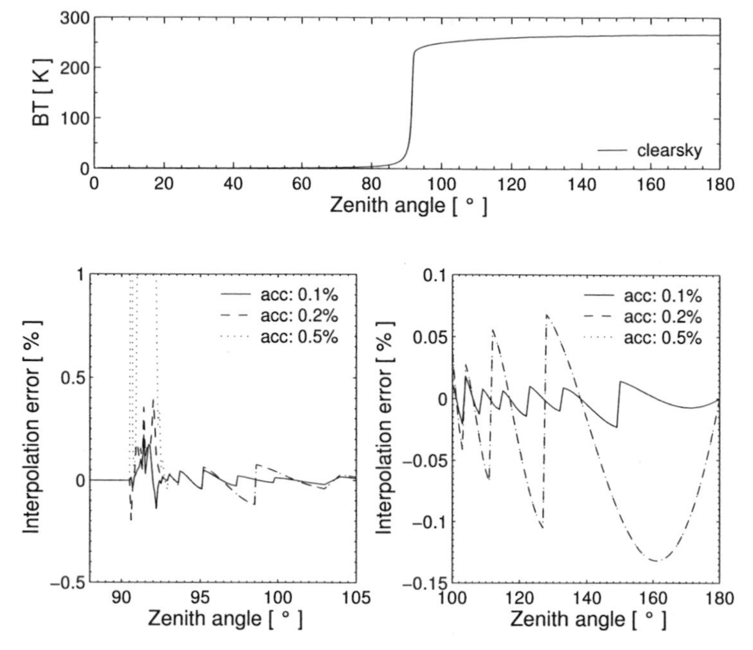

Figure 4.5: Interpolation errors for different grid accuracies. Top panel: Clear sky radiation simulated for a sensor at an altitude of 13 km for all viewing directions. Bottom left: Grid optimization accuracy for limb directions. Bottom right: Grid optimization accuracy for down-looking directions.

added to the reduced grid until the maximum difference is below a requested accuracy limit.

The top panel of Figure 4.5 shows the clear sky radiation in all viewing directions for a sensor located at 13 km altitude. This result was obtained with a switched-off cloud box. The difference between the clear sky part of the ARTS model and the scattering part is that in the clear sky part the radiative transfer calculations are done along the line of sight of the instrument whereas inside the cloud box the RT calculations are done as described in the previous section to obtain the full radiation field inside the cloud box. In the clear sky part the radiation field is not interpolated, therefore we can take the clear sky solution as the exact solution.

The interpolation error is the relative difference between the exact clear sky calculation (cloud box switched off) and the clear sky calculation with empty cloud box. The bottom panels of Figure 4.5 show the interpolation errors for zenith angle grids optimized with three different accuracy limits (0.1%, 0.2% and 0.5%.). The left plot shows the critical region close to 90°. For a grid optimization accuracy of 0.5% the interpolation error becomes very large, the maximum error is about 8%. For grid accuracies of 0.2% and 0.1% the maximum interpolation errors are about 0.4% and 0.2% respectively. However for most angles it is below 0.2%, for all three cases. For down-looking directions from 100° to 180° the interpolation error is at most 0.14% for grid accuracies of 0.2% and 0.5% and for a grid accuracy of 0.1% it is below 0.02%.

4.3.2 Interpolation methods

Two different interpolation methods can be chosen in ARTS for the interpolation of the radiation field in the zenith angle dimension: linear interpolation or three-point polynomial interpolation. The polynomial interpolation method produces more accurate results provided that the zenith angle grid is optimized appropriately. The linear interpolation method on the other hand is safer. If the zenith angle grid is not optimized for polynomial interpolation one should use the simpler linear interpolation method. Apart from the interpolation of the radiation field in the zenith angle dimension linear interpolation is used everywhere in the model. Figure 4.6 shows the interpolation errors for the different interpolation methods. Both calculations are performed on optimized zenith angle grids, for polynomial interpolation 65 grid points were required to achieve an accuracy of 0.1% and for linear interpolation 101 points were necessary to achieve the same accuracy. In the region of about 90° the interpolation errors are below 1.2% for linear interpolation and below 0.2% for polynomial interpolation. For the other down-looking directions the differences are below 0.08% for linear and below 0.02% for polynomial interpolation. It is obvious that polynomial interpolation gives more accurate results.

Another advantage is that the calculation is faster because less grid points are required, although the polynomial interpolation method itself is slower than the linear interpolation method. Nevertheless, we have implemented the polynomial interpolation method so far only in the 1D model. In the 3D model, the grid optimization needs to be done over the whole cloud box, where it is not obvious that one can save grid points. Applying the polynomial interpolation method using non-optimized grids can yield much larger interpolation errors than the linear interpolation method.

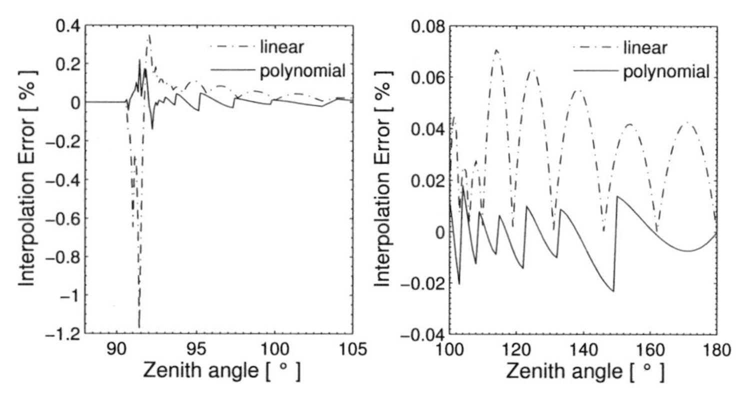

Figure 4.6: Interpolation errors for polynomial and linear interpolation.

4.3.3 Error estimates

The interpolation error for scattering calculations can be estimated by comparison of a scattering calculation performed on a very fine zenith angle grid (resolution 0.001° from 80° to 100°) with a scattering calculation performed on an optimized zenith angle grid with 0.1% accuracy. The cloud box used in previous test calculations is filled with spheroidal particles with an aspect ratio of 0.5 from 10 to 12 km altitude. The ice mass content is assumed to be $4.3 \cdot 10^{-3}\,\mathrm{g/m^3}$ at all pressure levels. An equal volume sphere radius of 75 µm is assumed. The particles are either completely randomly oriented (p20) or azimuthally randomly oriented (p30) (cf. Appendix 3.4). The top panels of Figure 4.7 show the interpolation errors of the intensity. For both particle orientations the interpolation error is in the same range as the error for the clear sky calculation, below 0.2 K. The bottom panels show the interpolation errors for Q. For the randomly oriented particles the error is below 0.5%. For the horizontally aligned particles with random azimuthal orientation it goes up to 2.5% for a zenith angle of about 91.5°. It is obvious that the interpolation error for Q must be larger than that for I because the grid optimization is accomplished using only the clear-sky field, where the polarization is zero. Only the limb directions about 90° are problematic, for other down-looking directions the interpolation error is below 0.2%.

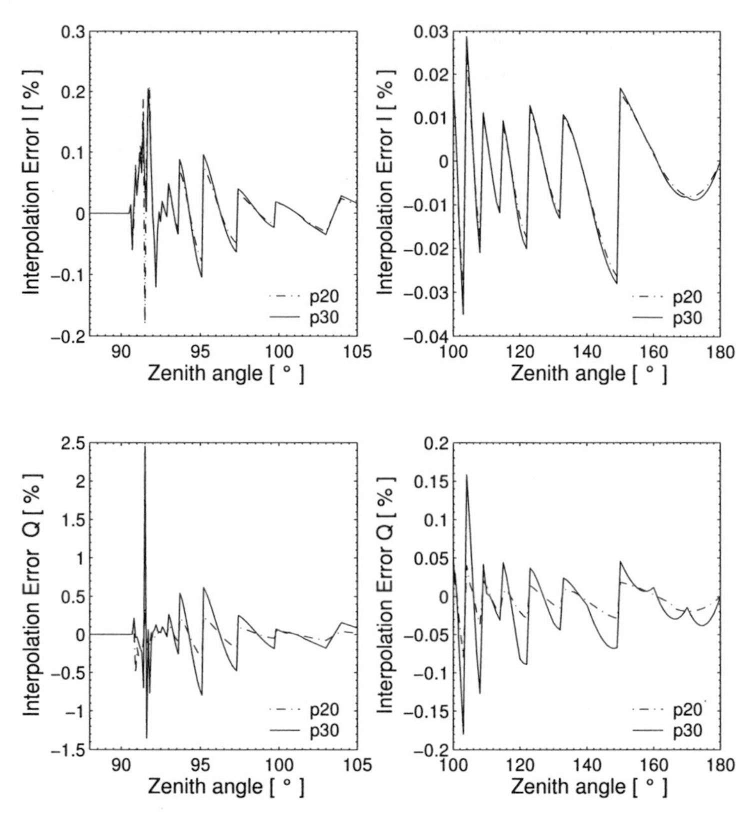

Figure 4.7: Interpolation errors for a scattering calculation. Left panels: Interpolation errors for limb directions. Right panels: Interpolation errors for down-looking directions. Top: Intensity I, Bottom: Polarization difference Q

5 Comparison of the DOIT method with other scattering models

5.1 Comparison with the pseudo-spherical unpolarized model FM2D

An intercomparison between the two-dimensional forward model FM2D developed at RAL (Rutherford Appleton Laboratory) and the ARTS-DOIT scattering model was performed in order to validate the FM2D model, which includes several approximations for efficiency reasons. The focus in the FM2D development was, that the model should be sufficiently fast for use in non-linear retrieval simulations.

5.1.1 The pseudo-spherical approach

The RAL forward model is briefly described and validated in von Clarmann et al. (2003). The model allows modeling of the atmosphere in two dimensions with the field expressed as a two-dimensional, vertical and horizontal, section. Operation in 1D mode, where the atmosphere is assumed to be spherically symmetric, is also possible. The model can be used for radiative transfer calculations in the microwave and in the mid-infrared wavelength regions. Scattering has been included by using the plane-parallel version of the GOMETRAN model (Rozanov et al., 1997), which was extended to include thermal emission as well as solar radiation. This allows to efficiently calculate the multiple-scattering source function at all points along the limb line-of-sight, ray-traced through a spherical atmosphere. The implementation of the scattering in FM2D is described in Kerridge et al. (2004), where the following intercomparison study has also been published.

Integration of the RTE

As the full solution of the VRTE (1.42) is computationally expensive, the following assumptions are included in FM2D for simplification:

1. Polarization can be neglected, i.e., the SRTE (1.51) is considered.
2. The scattering integral can be evaluated using a plane-parallel, 1D-model with optical properties taken from the 2D-atmosphere at the horizontal position of the considered propagation path point. Having obtained the scattering integral, which is often called source function, at all points along the line-of-sight, the scalar radiative transfer equation with a fixed scattering integral term Equation (4.20) is evaluated along the LOS.

The scattering integral is modeled using a modified version of the GOMETRAN++ forward model. The integration of the radiative transfer equation is done numerically by iterating the following equation, which follows from Equation (1.51), along the line of sight:

$$I_{l+1} = I_l e^{-\overline{\langle K_{11,l}\rangle}\Delta s_l} + \frac{\overline{\langle a_1, l\rangle}\,\bar{B} + \overline{\langle S_l\rangle}}{\overline{\langle K_{11,l}\rangle}}\left(1 - e^{-\overline{\langle K_{11,l}\rangle}\Delta s_l}\right), \quad (5.1)$$

where l is the index of a path segment and Δs_l the path length. The calculation of the source function $\langle S\rangle$ is done for all along track grid points as a first step. $\langle\overline{S_l}\rangle$ is then obtained by interpolating $\langle S\rangle$ in altitude, horizontal along-track position and local LOS zenith angle.

Representation of scatterers

The distribution of scatterers in FM2D is defined by the user as the 2D distributions of the associated scattering and absorption coefficient (in km^{-1}), together with parameters describing the phase function. The scattering properties are calculated outside FM2D using for example a Mie scattering program.

In principle the phase function must be specified in FM2D as a function of altitude, along-track position, frequency and scattering angle. In order to minimize the number of input parameters, the angular dependence of the phase function is modeled by two parameters: the Henry-Greenstein asymmetry parameter and the Rayleigh fraction.

Notes on accuracy of the model

For 1D geometries, the plane-parallel approach is well known to be accurate for near-nadir geometries. The extension to limb geometry in FM2D should give similar accurate results, since the plane parallel approximation is only made to determine the scattering integral, where an integration is performed over all spatial directions. Therefore the accuracy of the scattering integral will be similar for nadir and limb viewing geometry. This argument does not hold for strongly peaked phase functions which imply strong forward scattering. In this case, near-limb incident directions dominate the scattering integral and are not well modeled in the plane-parallel approach.

For 2D calculations many important aspects of the radiative transfer along the line of sight are captured by the approach since in most cases the local vertical atmospheric state profiles are dominant in the scattering integral calculation.

5.1.2 Clear sky comparison

Before comparing the scattering models it had to be assured that the clear sky calculations agree within a required accuracy.

In these calculations the same atmospheric profiles, gas absorption coefficients and general model settings were used. The MASTER-C band was chosen for the intercomparison as it includes different levels of gas absorption, a low gas absorption region at about 318 GHz and strong gas absorption at about 324 GHz.

The comparison is based on the standard profiles used at RAL. The species in the atmosphere are restricted to H_2O and O_3 and the earth is considered to be spherical with a radius of 6367.62 km. All atmospheric fields are defined on a 1 km vertical grid, irrespective of forward model internal grids, and extend to a height of 50 km. The assumed satellite altitude is 820 km. We have chosen a coarse optimized frequency grid including 70 frequencies. This is sufficient for an intercomparison of the scattering models, since the scattering properties do not change much in the MASTER band.

The results of the intercomparison, which are presented in Fig-

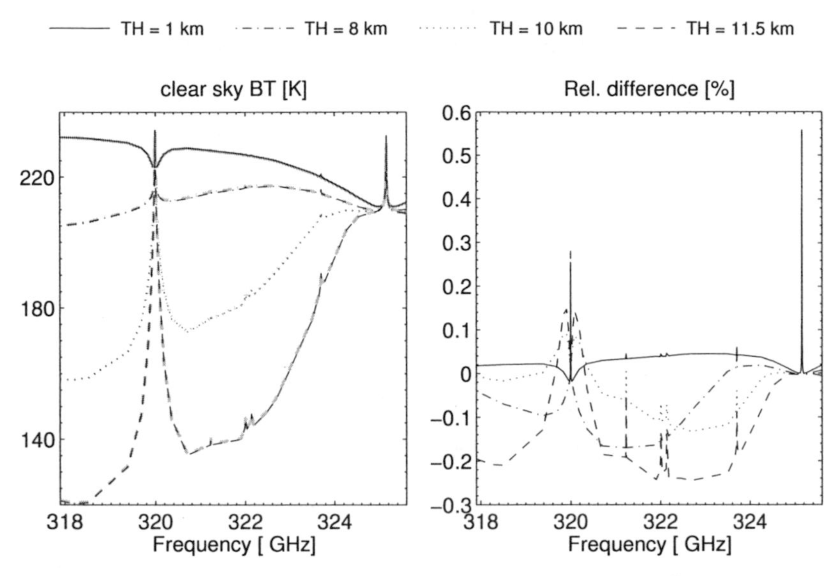

Figure 5.1: Clear sky radiative transfer comparison at different tangent heights (TH).

ure 5.1, show that the level of agreement between the two models is very good. For a propagation path step length of 1 km, the highest relative difference is about 0.6% for a non-refractive atmosphere, this corresponds to a difference of approximately 1 K in brightness temperature.

5.1.3 Comparison for cloudy scenarios

Definition of cloud scenarios

The cloud setup for the calculations is summarized in Table 5.1. The table includes the ice mass content IMC, the effective radius R_{eff} of the size distributions, and the altitude of the different cloud scenarios. The cloud particle size was parametrized according to the gamma distribution introduced in Section 3.5.2. All particles were assumed to be spherical. According to FIRE measurements, these scenarios correspond to realistic cloud cases.

Table 5.1: Definition of cloud scenarios for the scattering model intercomparison (DOIT versus FM2D)

Cloud	IMC [$g\,m^{-3}$]	R_{eff} [μm]	Altitude [km]
1	0.0001	21.5	10 – 12.1
2	0.004	34.0	10 – 12.1
3	0.02	68.5	10 – 12.1
4	0.04	85.5	10 – 12.1
5	0.1	128.5	10 – 12.1

Results and discussion

The result of the calculations is presented in Figure 5.2. The rows in the figure correspond to the five scenarios. The left column shows the simulated radiances at tangent heights of 1 km, 8 km, 10 km and 11.5 km. The black lines corresponds to the ARTS-DOIT results and the grey lines to the FM2D results. The middle column shows the scattering effect, which is the difference between cloudy radiances and clear sky radiances. A positive value corresponds to a brightness temperature enhancement due to cloud and a negative value corresponds to a brightness temperature depression. Again, the black lines are the ARTS results and the grey lines the FM2D results. The plots show, that the modeled radiances are very similar. The right column in the figure shows the differences between ARTS and FM2D, more precisely, the difference between the simulated scattering effects.

For scenario 1, the weakest cloud scenario, the difference between the models is below 0.004 K. The thicker the cloud the larger the cloud signal and also the difference between the models. The maximum difference for scenario 2 is approximately 0.3 K. For scenario 3 and 4 the maximum difference is about 1.6 K and for scenario 5 about 3.6 K. In scenarios 3 and 4 the FM2D results show larger brightness temperatures whereas in scenario 5 the ARTS brightness temperatures are higher. Figure 5.3 shows the normalized phase functions for the

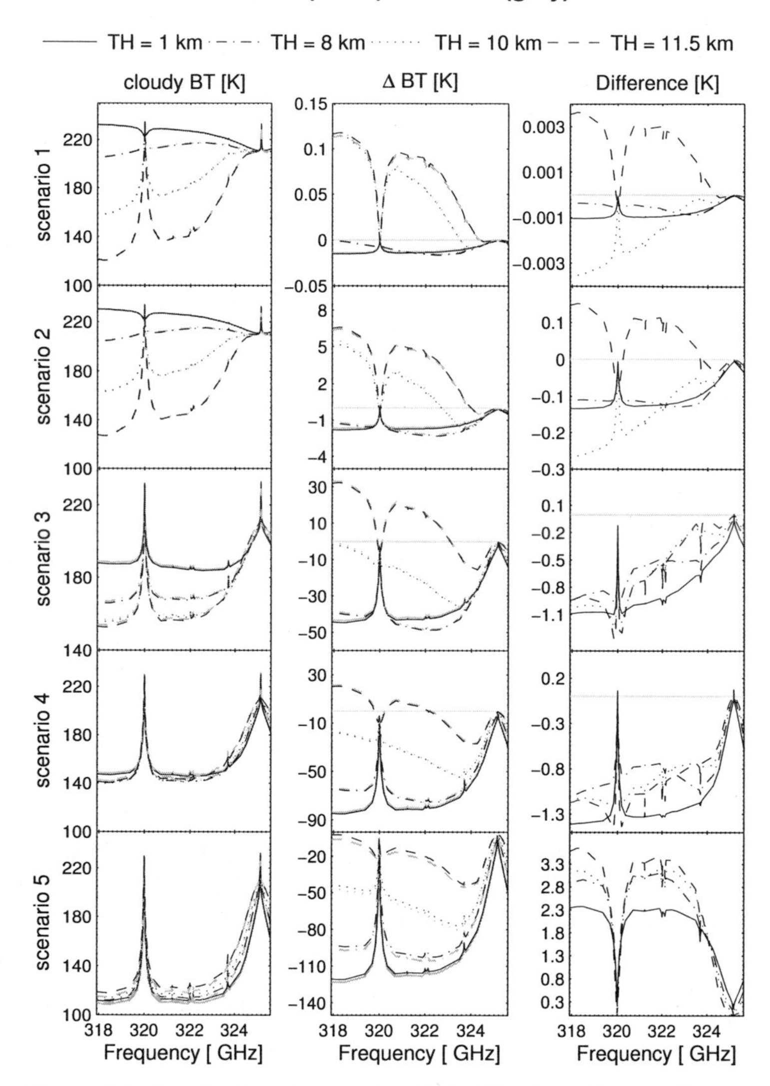

Figure 5.2: Results for all scenarios. Left: Cloudy radiances. Centre: Differences between cloudy and clear sky brightness temperatures. Right: Differences between ARTS and FM2D.

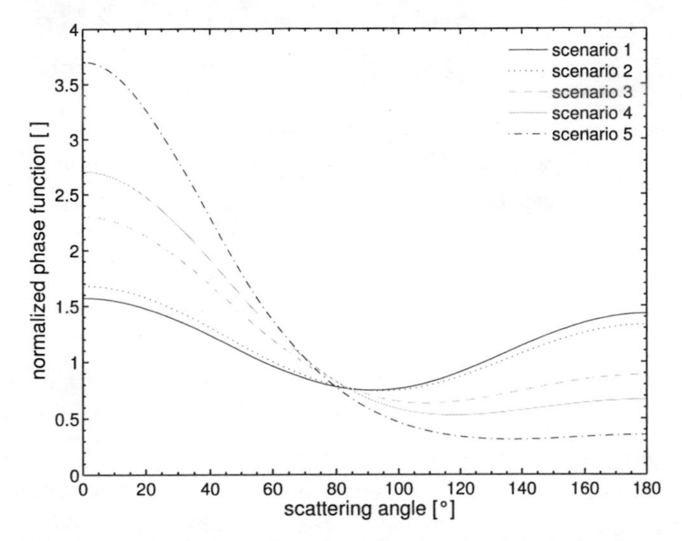

Figure 5.3: Normalized phase functions for the simulated cloud scenarios

modeled cloud scenarios. For scenario 1 and 2, the phase function shows a forward (0°) and a backward (180°) peak. The shape of the phase function is rather flat, that means radiation is scattered into all directions. For scenarios 3 to 5 most of the radiation is scattered into the forward directions. In scenario 5 the forward peak is most pronounced. The more radiation is scattered into forward direction, the less accurate is the plane parallel approximation for the calculation of the scattering integral. Both the increase in cloud optical depth and the more pronounced forward peak in the phase function explain the increasing differences between ARTS and RAL-FM2D with increasing ice mass content and increasing particle size. Simulations for similar cloud scenarios are also presented in Chapter 6, where the effect of different cloud parameters is discussed in detail using the ARTS-DOIT model.

5.2 Comparison with the single scattering model KOPRA for IR wavelengths

This section presents the intercomparison of the DOIT multiple scattering algorithm with the KOPRA (Karlsruhe Optimized and Precise Radiative Transfer Algorithm) model, which includes a single scattering approach. This intercomparison has been published in Hoepfner and Emde (2005). The validity of single scattering radiative transfer calculations for simulations of limb emission measurements of clouds in the mid-infrared spectral region was investigated. This study assesses the applicability of relatively fast single scattering calculations, which are important for data analysis of measurements of polar stratospheric and of cirrus clouds by current and future satellite borne spectrally high-resolution limb-emission sounders. Such instruments are for instance MIPAS (Michelson Interferometer for Passive Atmospheric Sounding) on Envisat (Fischer and Oelhaf, 1996), launched in March 2002, or TES (Troposhperic Emission Spectrometer) on EOS-Aura (Beer et al., 2001), launched in July 2004.

5.2.1 Zero- and single scattering solutions

KOPRA was especially developed for the analysis of spectrally high resolved remote sensing measurements of the earth's atmosphere in the mid-infrared (Stiller, 2000; Stiller et al., 2002). The part of the model describing radiative transfer in the gaseous atmosphere has been validated extensively (Glatthor et al., 1999; von Clarmann et al., 2002, 2003; Tjemkes et al., 2003). Based on a layer-by-layer approach KOPRA models the radiative transfer as a succession of extinction, emission and scattering in homogeneous layers.

From the analytic solution of Equation (1.51) for a homogeneous layer of thickness s with fixed scattering integral (4.21) we can derive the discretized radiative transfer equation. If the instrumental line-of-sight traverses L layers it reads:

$$I(s^{\mathrm{obs}}) = I^{(0)} \prod_{l=1}^{L} \tau(l) + \sum_{l=1}^{L} \left[\frac{\langle a_{1,l} \rangle \, \overline{B_l} + \langle S_l \rangle}{\langle K_{11,l} \rangle} (1 - \tau_l) \prod_{j=l+1}^{L} \tau_j \right], \tag{5.2}$$

where l the index of the layer and $\tau_l = \exp(- \langle K_{11,l} \rangle \, s_l)$ is the transmission of layer l with thickness s_l. For the determination of the scattering integral $S_l = \int_{4\pi} \mathrm{d}\boldsymbol{n}' \langle Z_{11,l}(\boldsymbol{n}, \boldsymbol{n}') \rangle I_l(\boldsymbol{n}')$, the incoming radiances are calculated neglecting the scattering source term:

$$I(s') = I^{(0)} \prod_{l'=1}^{L'} \tau(l') + \sum_{l'=1}^{L'} \left[\frac{\langle a_{1,l'} \rangle \, B_{l'}}{\langle K_{11,l'} \rangle} (1 - \tau_{l'}) \prod_{j=l'+1}^{L'} \tau_j \right]. \tag{5.3}$$

The prime symbol denotes that the variables belong to the first order scattering rays.

Below, four different options of scattering in KOPRA will be compared with the ARTS DOIT scattering algorithm:

KOPRA(0) Zero scattering scheme neglecting the scattering source term $\langle S_l \rangle$ in Equation (5.2).

KOPRA(1) Zero scattering scheme neglecting the scattering source term $\langle S_l(\boldsymbol{n}) \rangle$ and replacing the absorption coefficient $\langle a_{1,l} \rangle$ by the extinction coefficient $\langle K_{11,l} \rangle$ in Equation (5.2). This increases the thermal source term and might compensate for omitting the scattering source term. For optically thick clouds this approach should result in the blackbody radiation emitted from the top of the cloud, since extinction and emission are equal.

KOPRA(2) Single scattering scheme using Equation (5.2) and Equation (5.3).

KOPRA(3) Single scattering scheme using Equation (5.2) and Equation (5.3) in which $\langle a_{1,l'} \rangle$ is replaced by $\langle K_{11,l'} \rangle$, which could possibly compensate for neglecting the multiple scattered component of the radiation field.

5.2.2 Definition of scenarios

In order to avoid any problems with differences in line-by-line absorption calculations, absorption cross-sections were calculated using the KOPRA model. The spectral interval 946.149–950.837 cm^{-1} was selected and the gases CO_2 and H_2O were taken into account. The cross-sections were calculated for the atmospheric pressure-temperature profile on a 0.5 km grid. These pre-calculated cross-sections were used to simulate the gaseous radiance contributions with the ARTS model.

The altitude profile of the cloud was defined between 9.5 and 12.5 km altitude with linearly increasing (from 0 cm^{-3}) values of the particle number density from 9.5 to 10 km and decreasing values (to 0 cm^{-3}) from 12 to 12.5 km. Between 10 and 12 km the number density was constant. A log-normal size distribution of spherical ice-cloud particles was assumed with a median-radius of 4 µm and a width of 0.3. Table 5.2 summarizes the five scenarios of increasing density. Here, the smallest volume density is in the order of that of typical polar stratospheric clouds of type I containing a large fraction of HNO_3 and scenario 2 is representative of polar stratospheric clouds of type II consisting of ice particles.

For the determination of single scattering properties, Mie calculations were done on basis of refractive indexes of ice by Toon et al. (1994). In the middle of the defined spectral interval the refractive index is (1.07+0.17i) leading to an absorption cross-section of 5.1×10^{-7} cm^2 and a scattering cross-section of 1.6×10^{-7} cm^2. This results in a single scattering albedo $\omega_0 = 0.24$. Hence this is a case of relatively strong absorption, since the chosen wave-number region is situated at the edge of an ice absorption peak in the mid-IR with a maximum around 830 cm^{-1}. To cover also a case of strong scattering and weak absorption we used an index of refraction of (1.25+0.018i) for the same wave-number region. This resulted in absorption and scattering cross-sections of 1.1×10^{-7} cm^2 and 5.9×10^{-7} cm^2, respectively (ω_0 = 0.84). Figure 5.4 shows the single scattering albedo for spherical ice particles in the mid-infrared as a function of wave-number and particle radius. Obviously, the chosen values for ω_0 cover a large fraction of the overall variability.

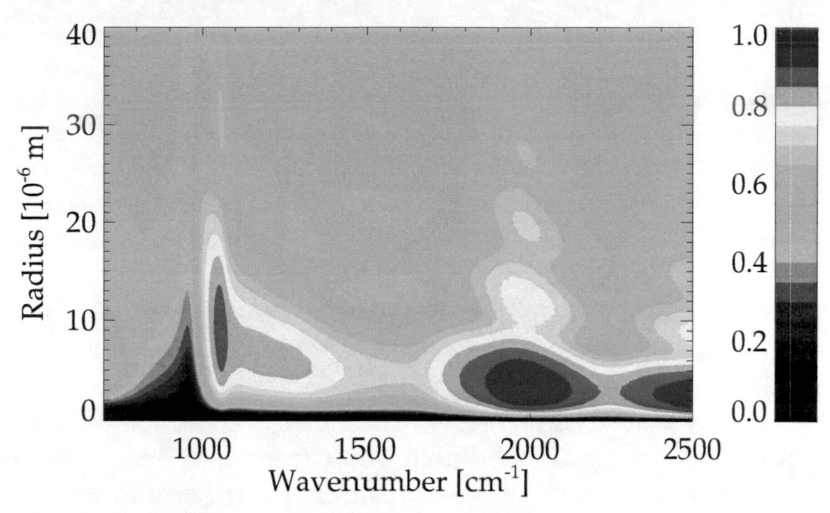

Figure 5.4: Single scattering albedo ω_0 for spherical ice particles of different radius covering the spectral region of the MIPAS/Envisat experiment. For the Mie calculations the refractive indexes of ice by Toon et al. Toon et al. (1994) were used.

Optical depths for the different cloud scenarios are given in Table 5.2 for nadir direction and limb view with a tangent altitude at 11 km in the middle of the cloud. In case of nadir geometry only scenario 5 is optically thick, while in limb direction scenarios 3–5 are opaque.

Table 5.2: Cloud scenarios used for the ARTS/KOPRA intercomparison. Optical depths are given for the two cases $\omega_0 = 0.24$ and $\omega_0 = 0.84$ (in brackets).

Cloud scenario	Number density [cm^{-3}]	Volume density [μm^3 cm^{-3}]	Optical depth nadir	Optical depth limb[a]
1	0.01	4.0194	$1.68(1.75)\times10^{-3}$	0.17(0.176)
2	0.1	40.194	$1.68(1.75)\times10^{-2}$	1.7(1.76)
3	1	401.94	$1.68(1.75)\times10^{-1}$	17.0(17.6)
4	10	4019.4	1.68(1.75)	170(176)
5	100	40194	16.8(17.5)	1700(1760)

[a] 11 km tangent altitude

5.2.3 Results for case $\omega_0 = 0.24$

Figure 5.5 shows the intercomparison between results from ARTS, and KOPRA(0)–KOPRA(3) for the case of strong absorption and for a line-of-sight crossing the middle of the cloud layer at 11 km tangent altitude. Spectra for lower altitudes are not shown here because the results were very similar. As reference, the top row compares the cloud-free model runs. In the region between the strong emission lines where the gaseous atmosphere is optically thin and to which, thus, the comparison between the cloud calculations will refer, differences are below 0.5%.

The very thin cloud of scenario 1 mainly introduces a broadband offset above the clear sky spectrum due to the emission by the cloud particles and the scattering of radiation from the troposphere and the earth's surface. The scattered contribution is the difference between KOPRA(0) and ARTS and accounts for about 35% of the total radiance. Since the second zero scattering scheme KOPRA(1) models a larger emission from the cloud particles, the difference to ARTS is with 15% somewhat reduced compared to KOPRA(0). The results of both single scattering models, KOPRA(2) and KOPRA(3) are nearly identical to the multiple scattering approach with less than 0.5% difference.

In scenario 2, the radiance continuum is strongly increased and reaches with 2000 nW/(cm^2 sr cm^{-1}) the value of the Planck function for the temperature at cloud altitude. The spectra of the single and multiple scattering models clearly show signs of radiance of tropospheric origin, like the downward pointing absorption features of the water vapor lines. These structures are missing in the calculations by both zero scattering schemes KOPRA(0) and KOPRA(1). The difference between those and ARTS are of the same magnitude as in scenario 1. The accuracies of KOPRA(2) and KOPRA(3) with less then 1% are still much better in case of neglecting scattering. However, with less than 0.5% KOPRA(3) fits closer to ARTS than KOPRA(2) with 1%.

In scenario 3 there is a further increase of the continuum radiance compared to scenario 2. The background value for the scattering mod-

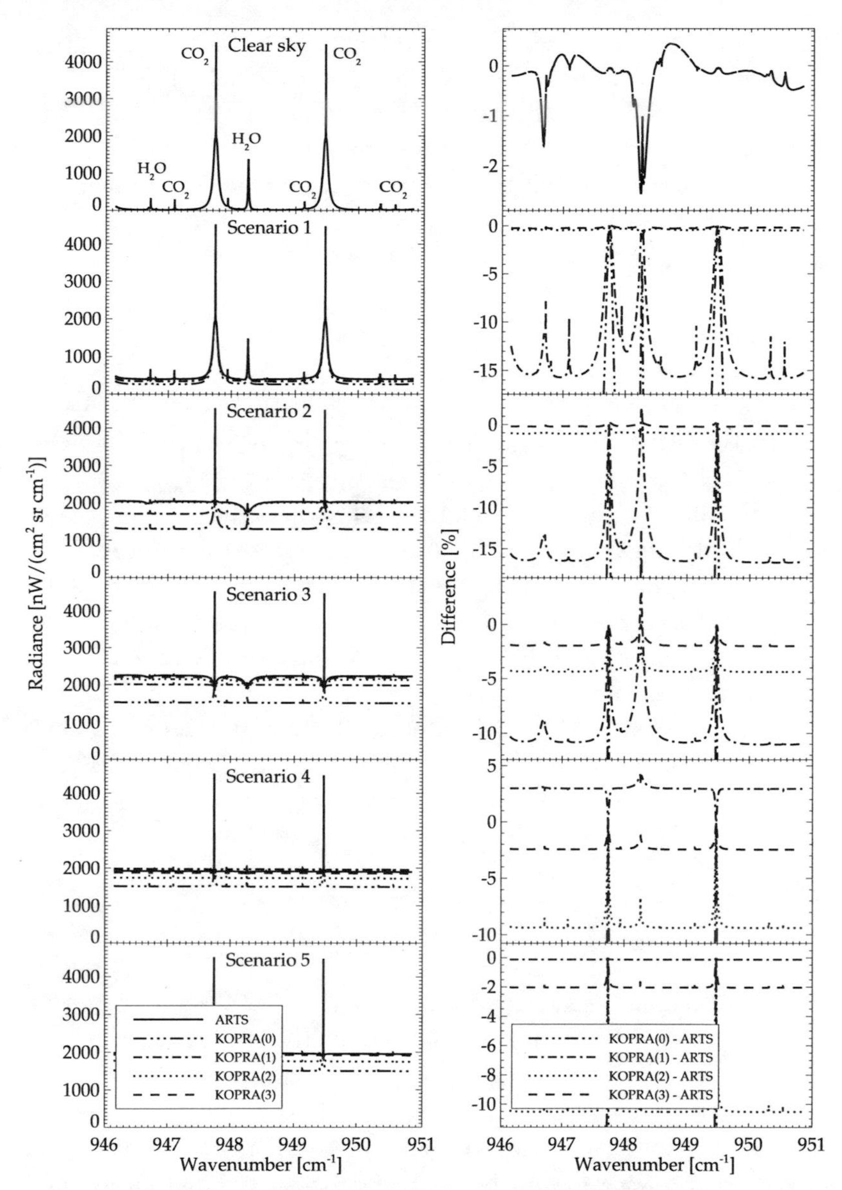

Figure 5.5: Comparison between limb spectra for 11 km tangent altitude calculated with ARTS and different KOPRA options for the case of strong absorption ($\omega_0 = 0.24$).

els is now larger than radiation of a blackbody at the cloud position which is given by the result of KOPRA(1). The cloud is optically thick in limb, but not in nadir direction. Thus radiance from the warm troposphere and the earth's surface increases the total signal. This is obvious from the fact that the spectra which include scattering, still exhibit the downward pointing features of the tropospheric water absorption lines. Differences show that the zero scattering models still underestimate the radiance by more than 10%. However, also the single scattering models differ from the ARTS reference by about 4.5% for KOPRA(2) and 2% for KOPRA(3).

The continuum radiance in scenario 4 is lower than in scenario 3 due to the fact that, though the cloud is still not opaque in nadir direction, less radiation from the troposphere reaches the particles along the line-of-sight which could scatter into the direction of the observer. Furthermore, the typical tropospheric absorption features are not visible anymore. KOPRA(1) is with 3% difference closer to ARTS than KOPRA(2) with about 9% difference. This is due to neglecting multiple scattering. KOPRA(3) with 2.5% deviation from ARTS still delivers the best result.

In case of scenario 5, which is optically thick in nadir and limb direction, the cloud closely resembles a black body. Thus, KOPRA(1) deviates from the multiple scattering model by only 0.1%. KOPRA(3) shows differences of 2% and KOPRA(2) of 11%.

5.2.4 Results for case $\omega_0 = 0.84$

Results of the model intercomparison for the case of large scattering ($\omega_0 = 0.84$) are shown in Figure 5.6. Though the optical depth of both cases is not very different (see Table 5.2) the continuum signal in the optically thin scenario 1 is increased by a factor of 1.4 compared to the case of strong absorption. This is due to the increased scattering of radiation originating in the warmer troposphere and on ground. Further, the flanks of the water vapor lines show downward pointing broader features also caused by scattered tropospheric radiation. With differences of 40% and more, the zero scattering models KOPRA(0) and KOPRA(1) are far off the reference while the single scattering schemes KOPRA(2) and KOPRA(3) deviate by only 1.5% and 1% from ARTS.

With about 3000 nW/(cm^2 sr cm^{-1}) the background radiance of the reference spectrum for scenario 2 is 50% higher than the blackbody radiance at cloud altitude. Very strong absorption features appear at the position of the water vapor and in the flanks of the CO_2 lines. Thus, as in scenario 1, zero scattering models are not capable to model these effects. However, compared to scenario 2 of the strong absorption case, even single scattering models show large differences compared to the ARTS reference: KOPRA(2) underestimates the radiance by about 7% while KOPRA(3) calculates 3% lower values than ARTS.

For scenario 3, which is optically thick in limb direction, the radiance calculated by ARTS reaches its maximum at about 3400 nW/(cm^{2}sr cm^{-1}). This is by a factor of 1.7 larger than the blackbody at cloud position. The single scattering model, however, does not follow this further increase, but results in lower radiances than in scenario 2. Thus, the differences with respect to the multiple scattering calculation increase up to 20% for KOPRA(3) and 35% for KOPRA(2).

For scenario 4, the ARTS radiances decreased and are only 1.2 times higher than the blackbody. Therefore, the zero scattering model KOPRA(1) compares best, with differences of 13%. The single scattering approach of KOPRA(3) is far off with up to 32% lower radiances.

For scenario 5, which is optically thick in limb and nadir directions, ARTS and blackbody calculations (KOPRA(1)) are nearly identical with less than 1% difference. Thus, quasi no radiation from the lower troposphere reaches the instrument any more. The comparison with KOPRA(3) is with 8% differences much better than for scenarios 3 and 4.

5.2.5 Summary and discussion

The validity range of zero and single scattering calculations for simulations of mid-IR limb emission measurements of clouds was investigated

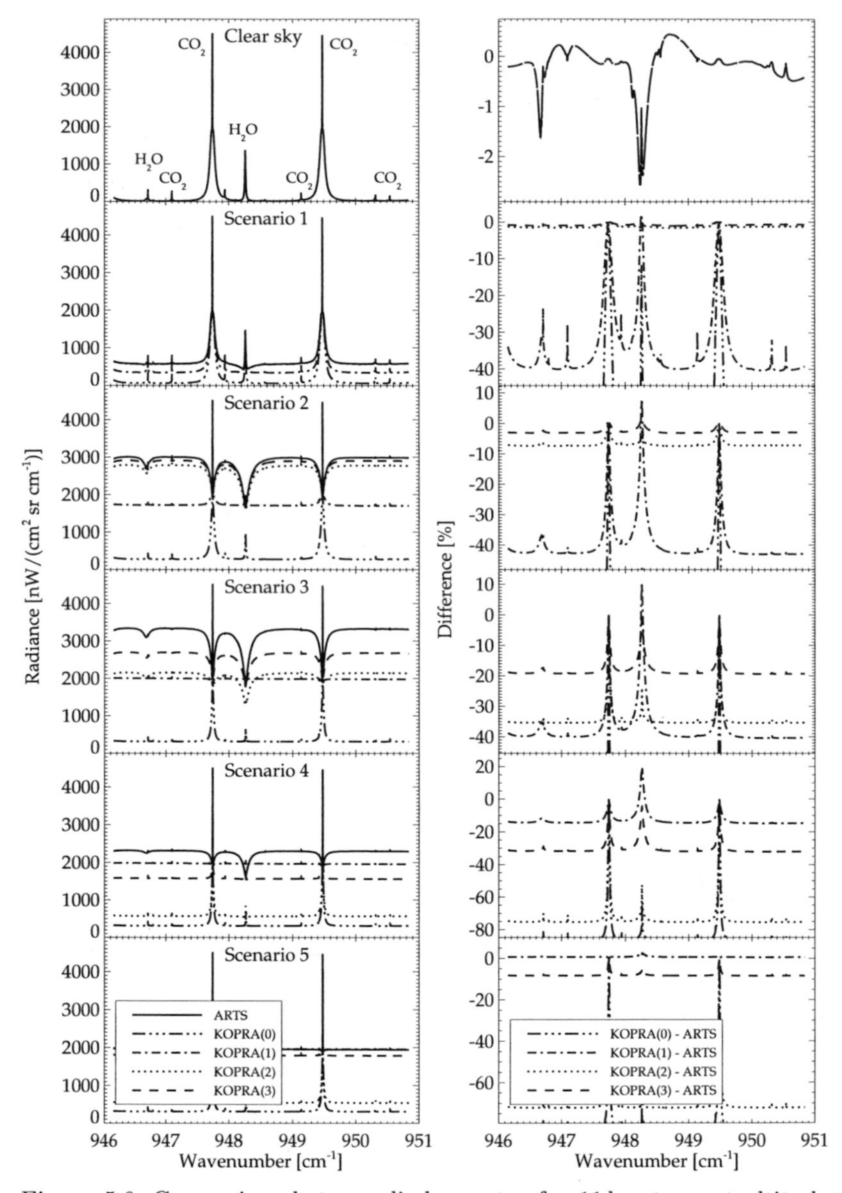

Figure 5.6: Comparison between limb spectra for 11 km tangent altitude calculated with ARTS and different KOPRA options for the case of strong scattering ($\omega_0 = 0.84$).

by comparison with the multiple scattering model ARTS-DOIT. Scenarios from optically thin to thick clouds with a low ($\omega_0 = 0.24$) and a high ($\omega_0 = 0.84$) single scattering albedo were used as baseline for the calculations.

For cloud scenarios 1 and 2, which are optically thin in limb direction, the single scattering approaches achieve results with maximum errors of a few percent. These cloud scenarios resemble polar stratospheric clouds and subvisible cirrus clouds. Thus single scattering models are sufficient for evaluation of such measurements. Zero scattering schemes, however, show errors of more than 15% and 40%, depending on the single scattering albedo. From Figure 5.4 it is clear that the zero scattering approaches can only be used for particle radii less then about 1 μm at the lower and less than 0.2 μm at the higher end of the shown spectral interval.

For clouds, which are optically thick in limb direction, the quality of single scattering calculations strongly depends on the single scattering albedo. The differences for the model KOPRA(3) range from only 2 – 3 % for $\omega_0 = 0.24$ up to 10 – 30% for $\omega_0 = 0.84$. For the latter case larger errors appear for clouds which are optically thick in limb, but not in nadir direction and which, thus, still scatter a large amount of lower tropospheric radiation into the instrumental line of sight. Combining these results with Figure 5.4 it is clear that for the regions with $\omega_0 > 0.8$, which are situated mainly above 1000 cm^{-1} for radii between 1 and 10–20 μm, the single scattering approaches deliver results with uncertainties in the range of the investigated high scattering case. However, in the atmospheric window below 1000 cm^{-1}, where the ice absorption peak is located, single scattering simulations in case of optically thick clouds should be reliable for particle sizes of less than 10–20 μm. For larger particles, the single scattering albedo is around 0.5 over the whole wavelength region. Here, we estimate that the accuracy of single scattering calculations lies between the extreme cases and is in the range of about 5 – 15%.

Comparing the results of different KOPRA options, we see, that KOPRA(1) and KORA(3) achieve better results than KOPRA(0) and KOPRA(2). The latter implementations underestimate radiances because they neglect the scattering source term while KOPRA(1) and

KOPRA(3) compensate for this by increasing the locally emitted radiation by replacing the absorption coefficient by the extinction coefficient in the direct and the scattered rays, respectively.

A further outcome of this study is that in cases with large single scattering albedo and clouds which are optically thick in limb, but thin in nadir direction, the continuum radiance can be by a factor of up to 1.7 enhanced with respect to the blackbody radiation at cloud top altitude. This is a consequence of the scattering of radiation from the warm troposphere and the earth's surface into the line-of-sight of a limb viewing instrument. Such strong effects are expected to be detectable in recently measured spectra by MIPAS on Envisat or in data of previous missions of mid-IR limb emission sounders like CRISTA (Cryogenic Spectrometers and Telescopes for the Atmosphere) (Spang et al., 2001) or CLEAS (Cryogenic Limb Array Etalon Spectrometer) (Roche et al., 1993).

5.3 Comparison between the Monte Carlo and the Discrete Ordinate approach

This section describes the comparison between the two ARTS internal scattering algorithms. In contrast to the previous sections, where the 1D DOIT algorithm was compared to other 1D scalar models, this section presents a comparison of 3D fully polarized models. The Monte Carlo and the DOIT models are the first models of this kind for radiative transfer modeling in the microwave region.

5.3.1 The Monte Carlo approach

A reversed Monte Carlo method has been implemented as a second scattering module besides the DOIT module into the ARTS model by Cory Davis. A strong consideration here was that the simplicity of the Monte Carlo radiative transfer concept should translate to reduced development time. Also, reversed Monte Carlo methods allow all computational effort to be concentrated on calculating radiances for the desired line of sight, and the nature of Monte Carlo algorithms makes parallel computing trivial.

Among the available Backward Monte Carlo RT models, several do not allow a thermal source, or do not consider polarization fully, which means that they can not handle a non-diagonal extinction matrix (e.g., Liu et al. (1996)). Some consider neither thermal source nor polarization (e.g., Oikarinen et al. (1999), Ishimoto and Masuda (2002)).

The ARTS Monte Carlo algorithm is described in Davis et al. (2004). The flowchart shown in Figure 5.7 illustrates the algorithm. In the following a short summary will be given without looking into the details:

1. Begin with a new photon at the cloud box exit point and sample a path length Δs along the first requested line of sight using the probability density function $g_0(\Delta s)$, which depends on the evolution operator $\tilde{\boldsymbol{O}}$ and on $\tilde{k}$, which is related to the extinction matrix.
2. A random number $\tilde{r}$ is drawn to choose between emission and scattering using a quantity similar to the scattering albedo, $\tilde{\omega}$.
3. If $\tilde{\omega} < \tilde{r}$, the event is considered to be emission, the reversed ray tracing is terminated and the Stokes vector contribution $\boldsymbol{I}^i$ is calculated using the Planck function $I_b(T)$, where T is the temperature. Then return to step 1.
4. If $\tilde{\omega} > \tilde{r}$, the event is considered to be scattering. At the scattering point a new incident direction $(\theta_{\text{inc}}, \phi_{\text{inc}})$ is sampled according to the probability function $g(\theta_{\text{inc}}, \phi_{\text{inc}})$, which is calculated using the phase matrix and the extinction matrix. The contribution to the intensity from this incident direction is obtained by the operator $\boldsymbol{Q}_k$.
5. A path length is sampled along the new direction.
6. If this path length leads the photon outside the cloud box the contribution of this photon on the cloud box boundary $\boldsymbol{I}^i$ is calculated using the matrix $\boldsymbol{Q}_k$.
7. Otherwise, if the sampled path length keeps the path within the cloud box, return to step 3.
8. When the Nth photon has reached the boundary of the cloud box the Monte Carlo algorithm is stopped and the field on the boundary

of the cloud box is used as the radiative background for the final cloud-sensor clear sky radiative transfer.

5.3.2 Setup

Atmosphere

The comparison simulations were performed for two frequencies, 122 and 230 GHz. These frequencies correspond to channels of the EOS MLS instrument, which can be used for cloud studies. Simulations for this instrument are shown in Chapter 8. Atmospheric profiles were taken from the FASCOD (Anderson et al., 1986) data for tropical regions. For 122 GHz, only the species O_2, N_2 and H_2O needed to be included. For 230 GHz, CO and O_3 were added. This selection of gaseous species is based on Waters et al. (1999). The water vapor profile was adjusted so that the relative humidity is 100% with respect to ice at altitudes with non-zero ice mass content.

Cloud scenario

A thin cirrus cloud layer was selected. The most simple case to start with is a box-shaped cloud in pressure, latitude and longitude coordinates. Since the Monte Carlo method is implemented only for 3D clouds, a 1D comparison is not possible. The FASCOD profiles however are 1D, so a box-shaped 3D cloud was embedded into the 1D atmosphere. The particle size distribution by Mc Farquhar and Heymsfield, which has been described in Section 3.5.3, was used. A single aspect ratio of 1.5 was applied for the whole cloud. It was assumed that the cloud particles are horizontally aligned. The ice mass content of the cloud was $0.1\,\mathrm{g/m^3}$ and the cloud altitude was 11.9 – 13.4 km. The horizontal extent was 400 km × 400 km, which corresponds to 3.6° latitude times 3.6° longitude for tropical regions. Two different sets of lines of sights were considered. They are illustrated in Figure 5.8. All lines of sight of set A intersect in the cloud at a point located 50 km away from the north edge. The lines of sight of set B intersect in the cloud in the center. Clear sky and cloudy simulations were performed

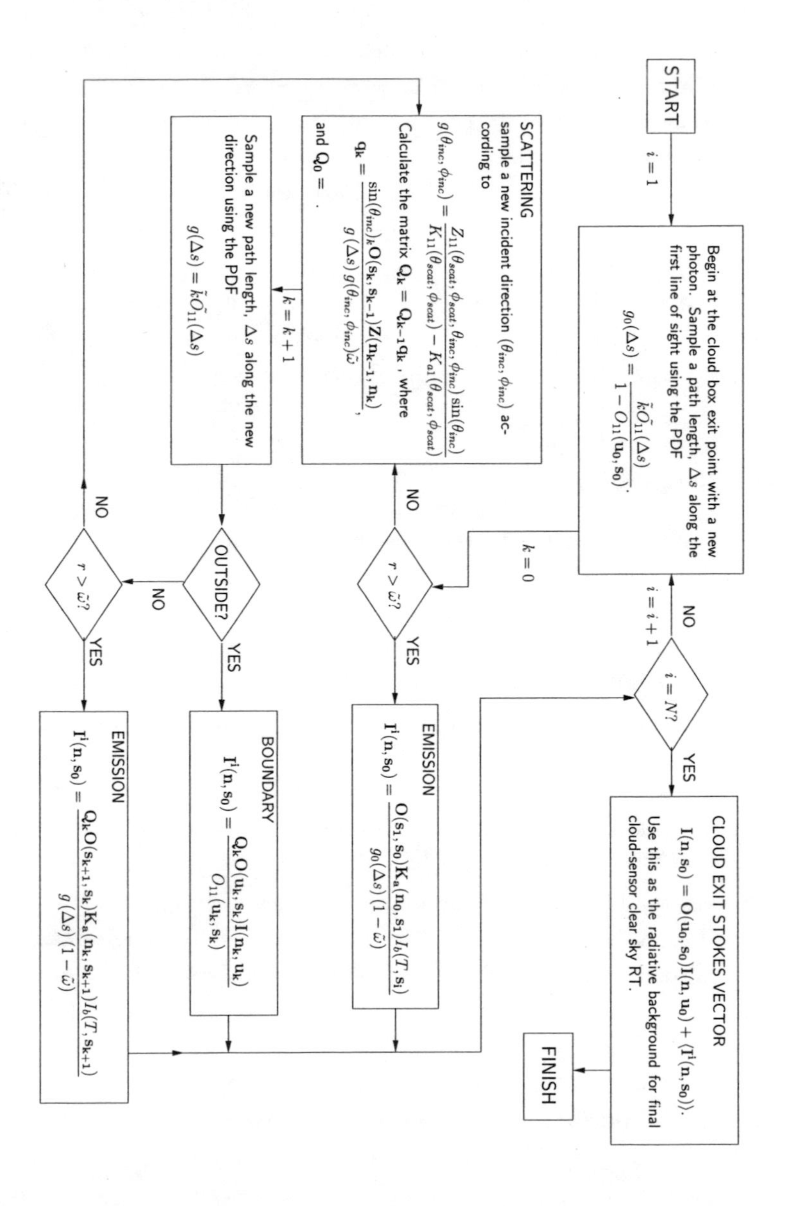

Figure 5.7: Flowchart illustrating the Monte Carlo algorithm. Courtesy of Cory Davis.

for tangent altitudes between 1 and 13 km. Since the DOIT model yields the full radiation field at the same time, a fine tangent altitude resolution was used (100 m). Monte Carlo simulations were performed on a tangent altitude grid with a resolution of 1 km. The zenith angle grid for the DOIT calculation was optimized to a grid accuracy of 0.1%. In 3D, so far only linear interpolation is implemented. This means that at critical zenith angles the error in the calculation can go up to approximately 1% (compare Figure 4.7). For most altitudes it is expected to be less than 0.2%. The error depends significantly on the cloud optical thickness. For very thick clouds it can be larger than the estimated result, since more iterations are required.

5.3.3 Results

The results for 122 GHz are presented in Figure 5.9. The top panel shows the obtained radiances in Rayleigh Jeans BT for tangent altitudes from 1 to 13 km. The grey line is the clear sky field. Black lines correspond to DOIT results and the markers are the Monte Carlo results for both LOS sets. In the radiance plot, there are no obvious differences between the DOIT and the Monte Carlo results or between the LOS set A and LOS set B. The middle left panel shows the difference between cloudy and clear sky radiances. Here we see that the Monte Carlo model shows less BT depression than the DOIT model. The absolute difference, which is shown in the bottom left panel, is between 0.2 and 0.4 K. The middle right panel shows the results obtained for the polarization difference Q, also in Rayleigh Jeans BT. Here also the Monte Carlo model shows slightly smaller values. The absolute difference between the models is approximately 0.05 K. For both, I and Q, the sign of the differences between the models is at 12 km tangent altitude opposite to the sign of the differences at other tangent altitudes. This can be explained by two opposing mechanisms: scattering away from the LOS for low tangent altitudes and scattering into the LOS for high tangent altitudes. If the scattering effect in the DOIT model was slightly greater than in the Monte Carlo model, the result would show more BT depression for low tangent altitudes and

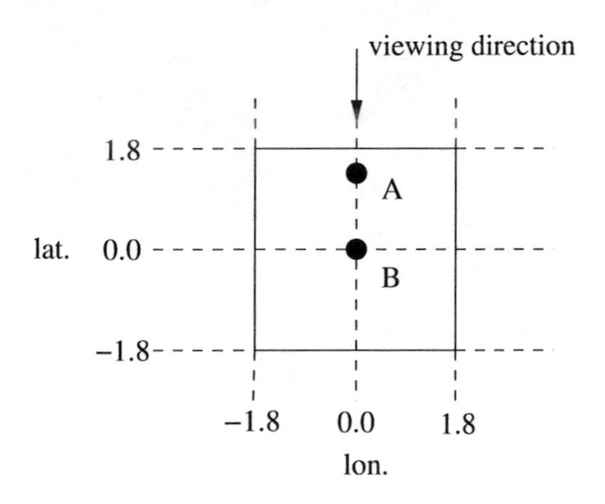

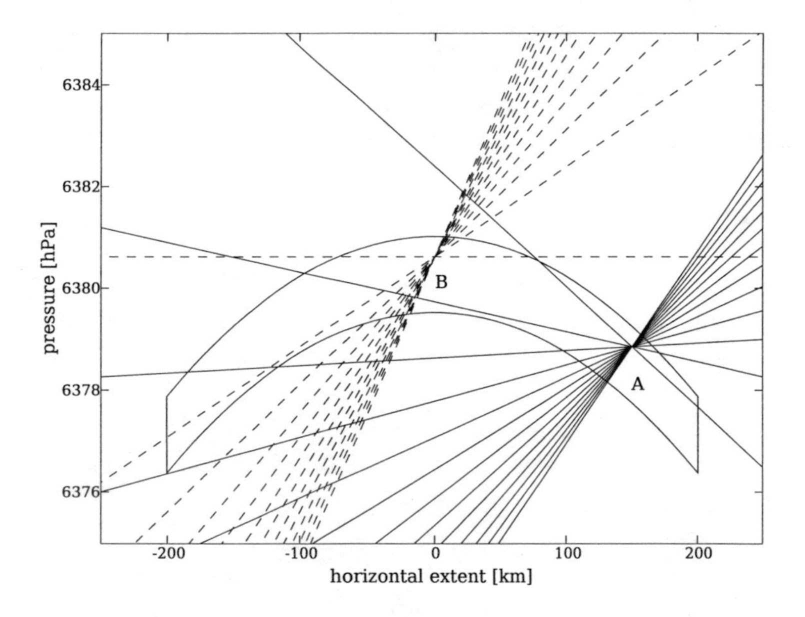

Figure 5.8: *Top:* Cloud box and crossing points of LOS sets A and B as seen from the top. The arrow shows the viewing direction of the instrument. *Bottom:* Lines of sight for tangent altitudes 1 − 13 km of sets A and B and the cloud box.

more BT enhancement for high tangent altitudes. At 12 km tangent altitude, scattering away from the LOS still dominates, but the BT depression becomes smaller. The middle left panel shows irregularities in the DOIT result due to interpolation. These would disappear with a finer zenith angle grid. The differences between the DOIT and the Monte Carlo models are similar for both sets of LOS, A and B. The agreement between the models is very good. It is well inside the error estimates of the DOIT model.

The results for 230 GHz are presented in Figure 5.10. The panels are arranged in the same way as in Figure 5.9. The radiance plot shows that the scattering effect is much larger for 230 GHz compared to 122 GHz. There is a BT depression of approximately 50 K for tangent altitudes below 8 km and a BT enhancement of up to 100 K for tangent altitudes above 8 km. At approximately 8 km there is almost no difference between cloudy and clear sky radiances. That means that at this point the same amount of radiation is scattered into the LOS as away from the LOS. The polarization difference is also much larger compared to 122 GHz, it goes up to 8 K. The middle panels show no obvious deviations between the Monte Carlo model and the DOIT model, but the bottom panels show, that the difference is almost 2 K for small tangent heights and 4 K at 12 km tangent altitude. For Q the absolute differences are small apart from 12 km tangent altitude where it goes up to -2 K. One of the reasons for this deviation is the interpolation of the radiation field. Since many iterations are required for the rather strong cloud the interpolation error could be more than 1%.

5.3.4 Discussion

At 122 GHz the Monte Carlo model and the 3D DOIT model are in very good agreement, the differences are less than 0.5 K for total intensities and about 0.05 K for the polarization difference. At 230 GHz the difference is much larger, for most tangent altitudes approximately 2 K for the intensities and approximately 0.3 K for the polarization difference. The major differences between the two frequencies are the gas

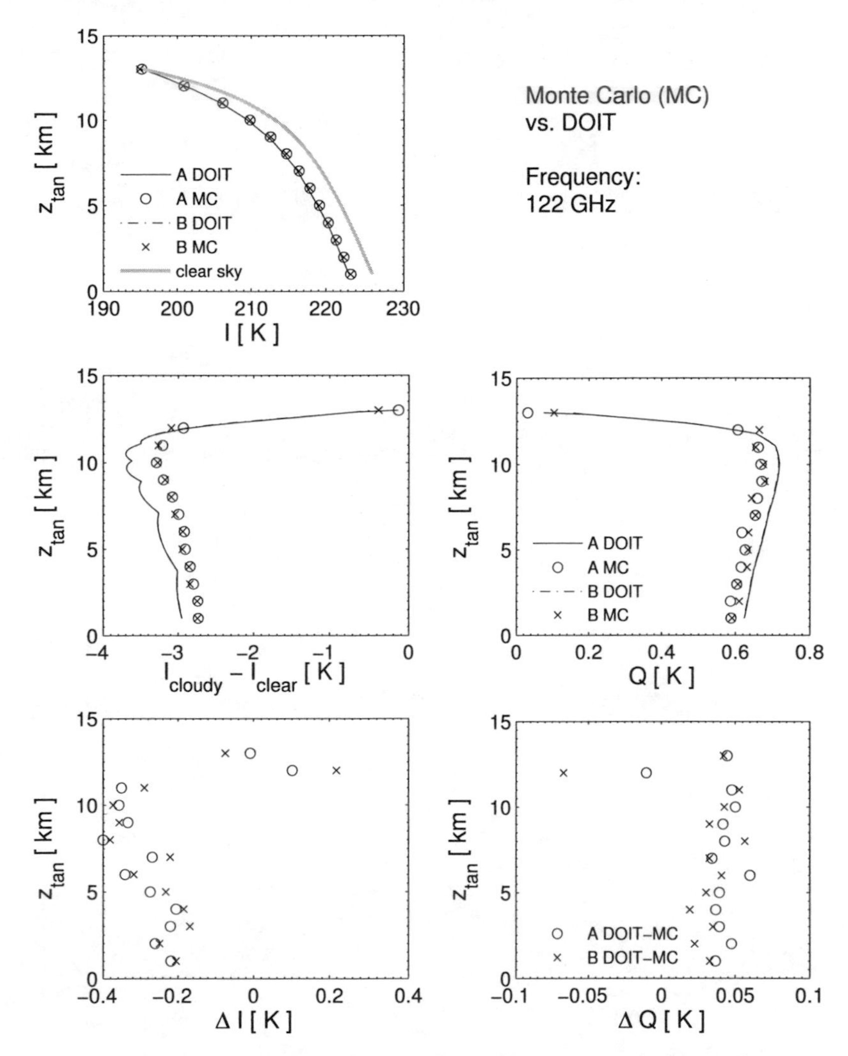

Figure 5.9: Results obtained for 122 GHz: The top panel shows the absolute radiances and the clear sky radiances (grey line). The middle left panel shows the intensity differences and the middle right panel shows the polarization signal. The bottom panels show the absolute differences between the DOIT and the MC modules for intensity and polarization.

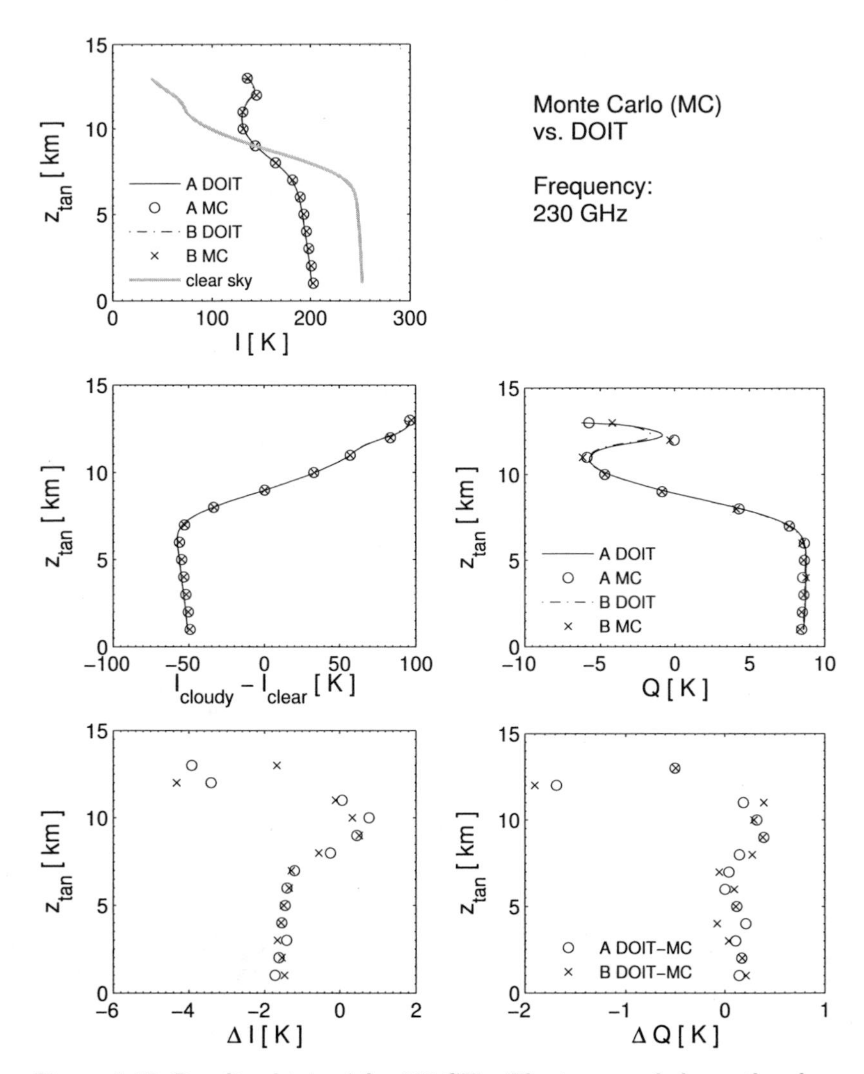

Figure 5.10: Results obtained for 230 GHz: The top panel shows the absolute radiances and the clear sky radiances (grey line). The middle left panel shows the intensity differences and the middle right panel shows the polarization signal. The bottom panels show the absolute differences between the DOIT and the Monte Carlo modules for intensity and polarization.

absorption and the single scattering properties. The clear sky field shows, that the gas absorption at high altitudes is much larger at 122 GHz compared to 230 GHz. The single scattering properties increase with frequency as can be seen from Figure 3.6. Therefore, cloud scattering has a much higher impact at 230 GHz.

The comparison shows, that both models are able to simulate the effect of cirrus clouds. The features of the scattering signal, i.e., only BT depression at 122 GHz and BT depression for low tangent altitudes and BT enhancement for high tangent altitudes at 230 GHz, are well brought out in the results. The numerical approaches to solve the VRTE Equation (1.42) are completely different, hence different numerical errors are involved. The accuracy of the Monte Carlo method is mainly restricted by the number of photons used for the calculations. In principle an arbitrary accuracy can be obtained by increasing the number of photons. The drawback is an increase in computation time.

A disadvantage of the 3D DOIT model is, that the numerical grids can not be chosen arbitrarily fine. The required memory for the computation depends on the size of the radiation field, which is discretized in pressure, latitude, longitude, incoming and scattered directions (compare Section 4.1.1). Also the computation time depends mainly on the discretization of the radiation field. Since the zenith angle grid is used for incoming and scattered directions, a doubling of the number of zenith angle grid points leads to four times greater computation memory and time requirements. A very fine zenith angle grid is unavoidable when one wants to achieve accurate results, since the radiation field strongly increases at zenith angels of approximately 90°. Doubling the size of the cloud box in all three spatial coordinates even leads to eight times greater computation time and memory requirements. The zenith angle grid discretization is not essential in the Monte Carlo model, since the radiation field is not interpolated in this dimension. Computation time also increases with the size of the cloud box, but it does not factorize like in the DOIT model. Another numerical error source is the calculation of 3D propagation paths. Propagation paths in a 3D atmosphere are calculated using an iterative approach, in contrast to 1D, where they are calculated analytically.

When the number of iterations increases, the numerical errors due to these calculations also increase.

Qualitatively the differences between the DOIT model and the Monte Carlo model can be understood, but the comparison shows that it is difficult to make correct error estimates. In both models, for 230 GHz, the estimated accuracy was better than the difference between the models, which means that there are additional numerical errors which need to be taken into account.

A feature of the DOIT model is that it yields the whole radiation field. This can be useful to obtain a basic physical understanding of for example polarization effects in the cloud. However, for limb sounding, when one is only interested in a few selected viewing directions, the Monte Carlo model should be preferred presently, if the cloud field is strongly inhomogeneous, so that a fine spatial discretization is required. In the future, with faster processors and more computation memory, one could also use the DOIT module for bigger cloud scenarios. For a thin cirrus layer with a large horizontal extent, it makes sense to use the 1D DOIT model, which is much faster than the Monte Carlo model.

5.3.5 Summary and conclusions

This comparison study was very important, since the two compared models are the very first ones, which can be used to simulate limb radiances in 3D spherical atmospheres for the microwave region. At the moment no other models are available for comparison. The result of this study is very promising. The agreement between the models is satisfactory. It shows that both the Monte Carlo method and the discrete ordinate method can be applied for solving the VRTE in a 3D spherical atmosphere. Optimization in accuracy and speed is planned to be implemented in the 3D-DOIT model in the future.

6 Unpolarized 1D simulations to study the impact of cirrus clouds on microwave limb measurements of the MASTER instrument

In this chapter the first simulations of microwave limb radiances with clouds are presented and analysed. They are computed using the 1D unpolarized version of the DOIT scattering module. Limb spectra are generated for the frequency bands of the MASTER (Millimeter Wave Acquisitions for Stratosphere/Troposphere Exchange Research) instrument (Buehler, 1999). The impact of various cloud parameters is investigated. Simulated brightness temperatures most strongly depend on particle size, ice mass content and cloud altitude. The impact of particle shape is much smaller, but still significant. Increasing the ice mass content has a similar effect as increasing the particle size, this complicates the prediction of the impact of clouds on microwave radiances without exact knowledge of these parameters. The work presented in this chapter has been published in Emde et al. (2004b).

6.1 General setup for the simulations

Before defining the cloud parameters the general setup of the model needs to be defined. This includes the composition and the geometry of the model atmosphere, the sensor and the numerical setup.

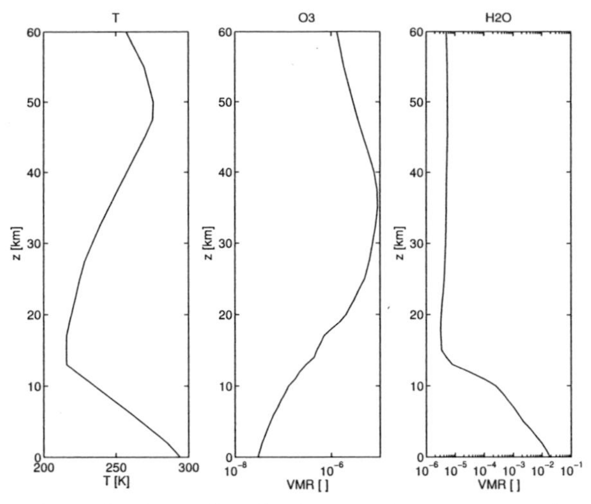

Figure 6.1: Atmospheric profiles for temperature, ozone and water vapor taken from FASCOD data for mid-latitudes in summer.

6.1.1 Atmosphere

The simulations are performed in a 1D spherical model atmosphere. Of course a 1D atmosphere is not a realistic environment for modeling clouds, because clouds are horizontally strongly inhomogeneous. Each cloud included in a 1D model corresponds to full cloud coverage around the globe. Thus the 1D calculations presented in this study can only be taken as an upper limit of scattering effects on the simulated radiances. Besides nitrogen and oxygen the two major atmospheric gases, water vapor and ozone, are included. The concentrations are taken from FASCOD (Anderson et al., 1986) data for mid-latitudes in summer. Profiles for temperature, ozone and water vapor are shown in Figure 6.1. Gas absorption is calculated based on the HITRAN (Rothman et al., 1998) molecular spectroscopic database using the ARTS model (first version ARTS-1-0). Refraction has been neglected in all calculations.

6.1.2 Sensor setup

The spectral ranges of bands B (293 – 306 GHz), C (317 – 326 GHz), D (342 – 349 GHz) and E (496 – 506 GHz) of the MASTER instrument are used for the simulations. The different scans correspond to tangent altitudes from 0 km up to 12.5 km, where 0 km altitude is the Earth's surface and 12.5 km is a tangent altitude above the cloud. For simplicity, the sensor is assumed to be ideal, which means that it measures exactly the intensity of the incoming radiation and is not subject to noise or other errors. This implies that the signal neither depends on the antenna pattern nor on the polarization of the measured radiation. The scalar version of the DOIT module is applied, thus only the first component of the Stokes vector is calculated, this means that polarization is neglected in the radiative transfer.

6.1.3 Numerical setup

For the accuracy of the results the discretization, especially in the angular domain, is very important. For the calculation of the scattering integral Equation (4.5), an equidistant zenith angle grid with an increment of 10° was taken. This is not sufficient for the radiative transfer calculation (Equation (4.7)), since the radiation field is strongly inhomogeneous around 90°. This problem and its solution are described in Section 4.3. For this study the zenith angle grid was optimized to represent the clear sky radiation field with an accuracy of 0.1%. The vertical grid is equidistant in altitude and the grid step-size is 0.5 km. If a propagation path step, i.e., the distance of two successive intersection points of the line of sight (LOS) with the vertical grid, is longer than 1 km, which occurs in limb geometry close to the tangent point, this step is divided into smaller (< 1 km) equidistant steps. The error of the results is estimated to be in the range of 0.5% including the interpolation of the radiation field as the main source for numerical errors.

6.2 Definition of cloud scenarios for the investigation of the impact of different cloud properties on limb radiances

To study the impact of different cloud properties (IMC, particle size, particle shape, altitude and frequency), the test cases compiled in Table 6.1 were defined. In all test cases, except case (3), it was assumed that the cloud consists only of spherical particles. Band C is taken for all calculations except for case (5). As the scattering signal depends also on the amount of gas absorption, it is interesting to study the effects in a band where we find frequency regions with high and others with low gas absorption (window regions). In this sense band C is the best selection.

Cloud height 10 – 12 km means that the scattering properties are defined on all pressure grid points between 10 and 12 km. The properties are linearly interpolated between the grid points, in this case they are interpolated between 9.5 and 10 km and between 12 and 12.5 km. Between 10 and 12 km the scattering properties are constant. the gamma distribution, which was described in Section 3.5.2, was used to calculate the single scattering properties and the particle number density fields.

Table 6.1 includes the following scenarios:

1. To investigate the impact of *particle size* spectra for the MASTER-C frequency band were calculated. The IMC is constant in all calculations ($1.6{\cdot}10^{-3}$ g/m^3) and the mean effective radius varies from 21.5 to 128.5 μm.
2. The influence of *cloud altitude* is also studied using band C. IMC and effective radius are constant, $1.6{\cdot}10^{-3}$ g/m^3 and 34.0 μm respectively. Calculations are performed for three different cloud altitudes: 6 – 8 km, 8 – 10 km and 10 – 12 km.
3. This is the only case were the cloud is assumed to consist of cylindrical particles. Spectra for five different aspect ratios in the range from 0.3 to 4.0 are calculated. The aspect ratio is the diameter of the cylinder divided by its length. Again band C is used and

IMC and effective radius are constant, $1.6{\cdot}10^{-3}\,g/m^3$ and $68.5\,\mu m$ respectively.

4. Measurements have shown that there is a *correlation between particle size and ice mass content.* Realistic scenarios according to a plot shown in Evans et al. (1998), which includes results from the FIRE (Kinne et al., 1997) campaign, were picked out. The IMC is varied from $4{\cdot}10^{-5}$ to $0.04\,g/m^3$ and the effective radius from 21.5 to $128.5\,\mu m$. Like in the other cases band C is taken and the cloud altitude is $10-12\,km$.
5. Calculations for different *frequency bands*, B, C, D and E are performed to see the impact of the same cloud in different frequency regions. IMC and effective radius are constant, $1.6{\cdot}10^{-3}\,g/m^3$ and $34.0\,\mu m$ respectively.

6.3 Results

6.3.1 Impact of particle size

Simulations for varying particle sizes and constant IMC are shown in Figure 6.2. The simulated radiances are presented in Rayleigh-Jeans brightness temperature (BT) units. On the left hand side we see the results at 8 km tangent altitude. The spectrum of the cloud consisting of the smallest particles ($R_{\text{eff}} = 21.5\,\mu m$) is almost identical to the clear sky spectrum. All other clouds show a BT depression in the window regions. For a particle size of $34.0\,\mu m$ the brightness temperature depression (ΔBT) is about 4 K. It becomes much larger with increasing particle size. For a particle size of $68.5\,\mu m$, ΔBT is approximately 17 K, for $85.5\,\mu m$ approximately 28 K and for $128.5\,\mu m$ approximately 52 K.

The plots on the right hand side show the results at 11.5 km tangent altitude, which look very different from those at 8 km tangent altitude. In the window regions about 317 and 322.5 GHz a BT enhancement is observed. This means, that more radiation is scattered into the line of sight (LOS) than away from the LOS. The largest BT enhancement can be observed at about 318 GHz in the window region of band C.

Table 6.1: Definition of five test cases to study the effect of cloud properties on limb radiances

	IMC [g/m^3]	R_{eff} [µm]	cloud alt. [km]	band	aspect ratio
1	$1.6\cdot10^{-3}$	21.5 34.0 68.5 85.5 128.5	10 – 12	C	—
2	$1.6\cdot10^{-3}$	34.0	6 – 8 8 – 10 10 – 12	C	—
3	$1.6\cdot10^{-3}$	68.5	10 – 12	C	0.3, 0.5, 1.0, 2.0, 4.0
4	$4\cdot10^{-5}$ $1.6\cdot10^{-3}$ $8\cdot10^{-3}$ 0.016 0.04	21.5 34.0 68.5 85.5 128.5	10 – 12	C	—
5	$1.6\cdot10^{-3}$	34.0	10 – 12	B, C, D, E	—

It ranges from about 5 K for a particle size of 21.5 µm and goes up to about 50 K for a particle size of 128.5 µm. Above approximately 322.5 GHz we see a BT depression for all particle sizes. The total extinction coefficient consists of particle absorption, particle scattering away from the LOS and gas absorption. If the contribution from the gas absorption is dominant, the scattering effect becomes small. More radiation is absorbed than scattered into the LOS. The BT depression varies from almost 0 K for the cloud with the smallest particles to approximately 35 K for the largest particles. The results show, that the size of the particles has a very large impact on limb radiances.

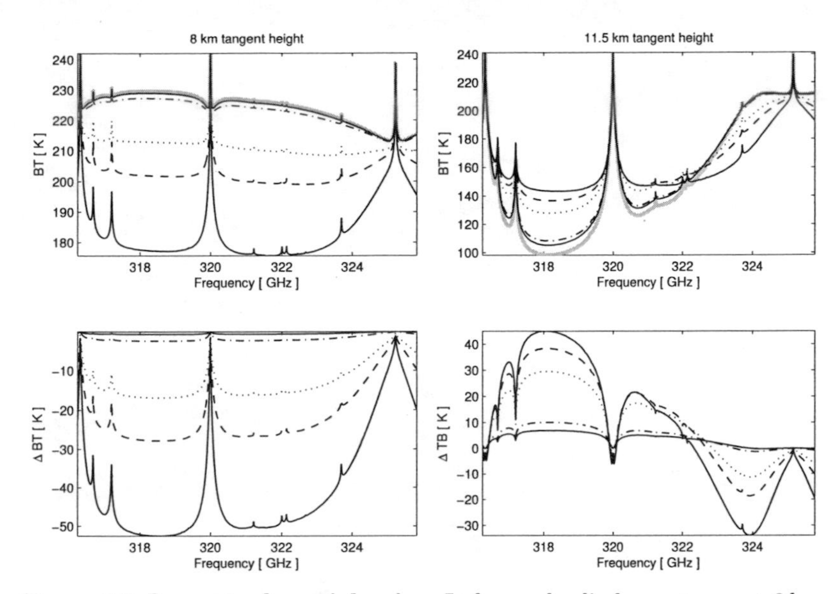

Figure 6.2: **Impact of particle size.** Left panel – limb spectrum at 8 km tangent altitude. Right panel – limb spectrum at 11.5 km tangent altitude. Clear sky spectrum (grey), and cloudy spectra for $R_{\text{eff}} = 21.5\,\mu\text{m}$ (—), $R_{\text{eff}} = 34.0\,\mu\text{m}$ (– · –), $R_{\text{eff}} = 68.5\,\mu\text{m}$ (· · ·), $R_{\text{eff}} = 85.5\,\mu\text{m}$ (– – –) and $R_{\text{eff}} = 128.5\,\mu\text{m}$ (—). The top plots show absolute BTs, the bottom plots show the differences between cloudy and clear sky spectra.

6.3.2 Impact of cloud altitude

The impact of cloud altitude on limb radiances is presented in Figure 6.3. The IMC was the same as taken for Figure 6.2. The assumed particle size for those calculations was 34.0 µm. This is one of the optically thinner clouds. The grey line is the clear sky spectrum. We see that the cloud at altitude 6 – 8 km has a very small impact on the radiances, below 0.5 K at 318 GHz and even less at higher frequencies. The cloud at 8 – 10 km altitude leads to a BT depression of maximal 1.7 K and the one at 10 – 12 km to a BT depression of maximal 2.4 K.

On the right side the BT is plotted as a function of tangent altitude. The lowest cloud shows in all tangent altitudes only a very small impact. The cloud at 8 – 10 km altitude leads to a BT depression at tangent altitudes up to 9 km and then to small BT enhancement up to

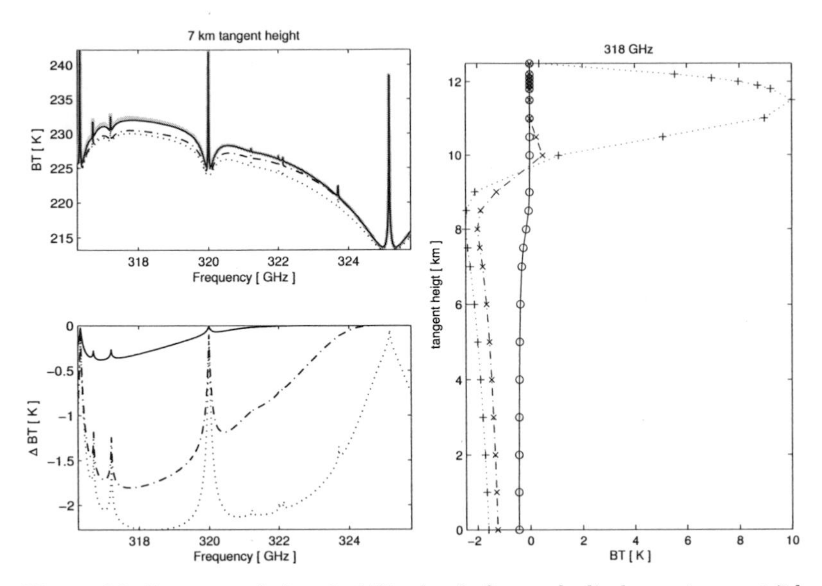

Figure 6.3: **Impact of cloud altitude.** Left panel - limb spectrum at 7 km tangent altitude. Clear sky spectrum (grey), and cloudy spectra for cloud altitude 6 – 8 km (—), 8 – 10 km (– · –) and 10 – 12 km (· · ·). Top plot shows absolute BTs, bottom plot differences from the clear sky case. Right panel - radiances at 318 GHz as a function of cloud altitude. 6 – 8 km (—), 8 – 10 km (– · –) and 10 – 12 km (· · ·).

10.5 km tangent altitude. The cloud signal is observed up to 10.5 km, since the scattering properties are interpolated linearly between the grid points. As mentioned above, the cloud ranges from 7.5 to 10.5 km in this case. For the highest cloud, we also observe a BT depression up to 9 km tangent altitude and a very high BT enhancement from 10 to 12 km tangent altitude. The enormously higher BT enhancement for the high cloud is due to the fact that in lower altitudes water vapor absorption is very large.

6.3.3 Impact of particle shape

The results of limb spectra for clouds consisting of cylindrical particles with different aspect ratios varying from 0.3 to 4.0 is shown in Figure 6.4. The calculations were done for the same IMC as taken in the

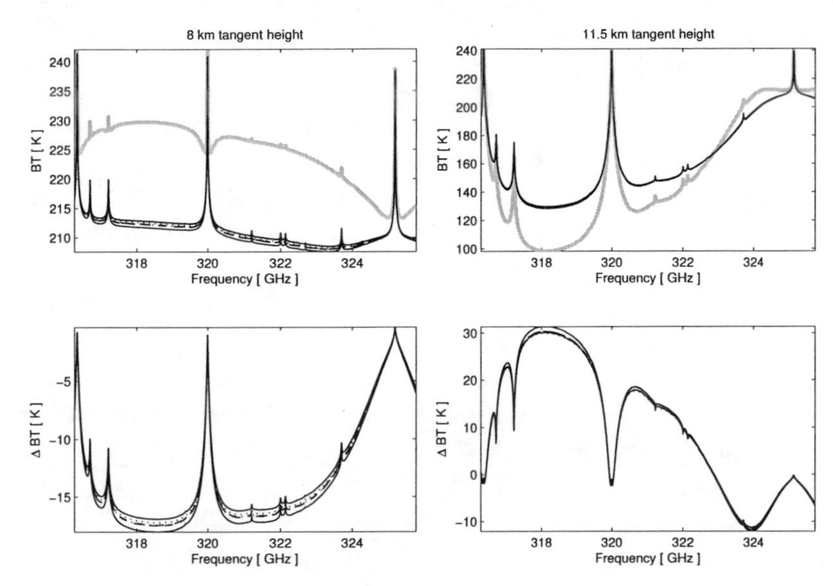

Figure 6.4: **Impact of particle shape.** Left panel - limb spectrum at 8 km tangent altitude. Right panel - limb spectrum at 11.5 km tangent altitude. Clear sky spectrum (grey), and cloudy spectra for aspect ratios 0.3 (—), 0.5 (– · –), 1.0 (· · ·), 2.0 (– – –) and 4.0 (—). The top plots show absolute BTs, the bottom plots show the differences between cloudy and clear sky spectra.

previous calculations, but here a larger mean particle size of 68.5 μm was taken. We can see immediately, that the difference between the curves is small compared to the scattering effect itself. At 8 km tangent altitude the total BT depression is about 17 K. The difference between the spectra for different aspect ratios is only approximately 2 K. At 11.5 km tangent altitude the maximal BT enhancement is about 30 K and the differences for different shapes are again in the order of 2 K. At this point it does not make sense to interpret the results in detail, for example, whether plates (aspect ratio > 1) show a higher signal than cylinders (aspect ratio < 1). It may only be concluded, that the effect of particle shape is much smaller than the effects of particle size and cloud altitude.

6.3.4 Cloud scenarios with correlation between IMC and R_{eff}

Results of the calculations for the "realistic" clouds are shown in Figure 6.5. The left side shows the results for 8 km tangent altitude. The impact of the optical thinnest cloud (IMC $= 4\cdot10^{-5}$ g/m^3, $R_{\text{eff}} = 21.5\,\mu$m) is very small, the spectrum can not be distinguished from the clear sky spectrum. The impact of the second cloud (IMC $= 1.6\cdot10^{-3}$ g/m^3, $R_{\text{eff}} = 34.0\,\mu$m), which is the one used for studying the effect of cloud altitude is very small compared to the other cloud scenarios. For the optically thickest cloud the BT depression may go up to 120 K. The highest BT enhancement at 11.5 km is observed for a medium cloud thickness (IMC $= 8\cdot10^{-3}$ g/m^3, $R_{\text{eff}} = 68.5\,\mu$m). The reason is that in case of very thick clouds, mainly radiation scattered into the LOS in the upper part of the cloud contributes to the spectra. It is very probable that radiation scattered into the LOS in the lower or middle part of the cloud is again absorbed or scattered away from the LOS by a cloud particle. For thinner clouds, most radiation, which is scattered once into the LOS, will continue to propagate into this direction without being disturbed by other cloud particles.

Although the micro-physical cloud properties, i.e., particle size and IMC are "realistic", a limb instrument would observe smaller scattering signals. As mentioned above, these calculations can only be taken as an upper limit of scattering effect, because the 1D mode of the model was used, assuming a homogeneous cloud cover. This assumption is unrealistic, particularly for the intense cloud cases, which correspond to clouds of limited horizontal extent.

6.3.5 Spectra for different frequency bands

In Figures 6.6 to 6.9 results for MASTER bands B, C, D and E are presented. They correspond to cloud case 2. The spectra are shown for the tangent altitudes of 3, 9, 10.5 and 11.5 km. The top panels show the clear sky spectra, the middle panels the spectra in the presence of clouds and the bottom panels show the difference ΔBT between

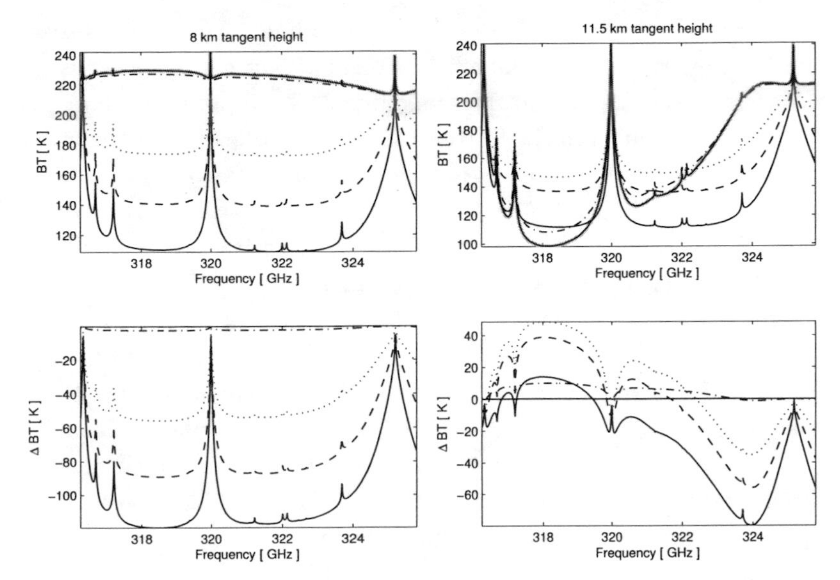

Figure 6.5: **Impact of** IMC. Left panel - limb spectrum at 8 km tangent altitude. Right panel - limb spectrum at 11.5 km tangent altitude. Clear sky spectrum (grey), and cloudy spectra for IMC = $4{\cdot}10^{-5}\,\mathrm{g/m^3}$ (—), $1.6{\cdot}10^{-3}\,\mathrm{g/m^3}$ (- · -), $8{\cdot}10^{-3}\,\mathrm{g/m^3}$ (· · ·), $0.016\,\mathrm{g/m^3}$ (– – –) and $0.04\,\mathrm{g/m^3}$ (—) and the corresponding particle sizes. The top plots show absolute BTs, the bottom plots show the differences between cloudy and clear sky spectra.

cloudy and clear sky spectra. IMC, R_{eff} and cloud altitude are defined in Table 6.1.

In band B there is an oxygen line at approximately 298.5 GHz. The other visible spectral lines are due to ozone. For 3 km tangent altitude there is a BT depression of maximal 1.5 K. All other tangent altitudes show a BT enhancement due to clouds. The difference between cloudy and clear sky spectrum is very small (< 3 K) for a tangent altitude of 9 km. This tangent altitude is below the cloud. As the considered cloud is rather optically thin, we can already see at 9 km a BT enhancement. For 10.5 and 11.5 km tangent altitude the BT enhancement can be more than 15 K.

In band C there are ozone lines around 317 and 320 GHz. The line at approximately 325 GHz is a water vapor line with a high absorption. This means that the largest scattering effect can be observed in the window regions around the ozone line at 320 GHz. Like in band B we see a BT enhancement for the tangent altitudes inside the cloud (10.5 km and 11.5 km). The maximum at 11.5 km tangent altitude for a frequency of about 318 GHz is with a brightness temperature difference of 10 K smaller than the maximum in band B, because the total gas absorption in band C is higher than in band B. The scattering coefficient increases with frequency in the microwave range (cf. Figure 3.6). At 11.5 km tangent altitude (upper part of the cloud) most radiation is scattered into the LOS. At 10.5 km tangent altitude, ΔBT is smaller compared to band B. As the scattering coefficient is larger in band C, the probability for multiple scattering is increased. If radiation is scattered into the LOS at 10.5 km there is a large probability that it is again scattered out of the LOS during the propagation from 10.5 to 12 km. The spectrum at tangent altitude 9 km shows already a BT depression in band C, the amount is less than 2 K. The BT depression at 3 km tangent altitude is maximal 2.5 K, about 1 K larger than the maximal depression in band B. This is the expected result, as the extinction coefficient is increased for higher frequencies.

In band D there are ozone lines around 343.3 GHz and an oxygen line at approximately 345.3 GHz. The total gas absorption is similar to band B. Since the scattering coefficient is larger for higher frequencies, the BT enhancement from the cloud at 10.5 and 11.5 km tangent altitude is increased, it is in this case maximal 11 K and 14 K respectively. At 9 km tangent altitude, the cloud effect is very small, only a depression of less than 1 K is observed. The BT depression at 3 km tangent altitude is approximately 3 K.

Band E is in a much higher frequency region with much larger scattering coefficients than the other bands. But also the total gas absorption is the largest in this band. Only for a tangent altitude of 11.5 km we see a BT enhancement throughout the whole band, the maximum value is about 17 K. The smallest brightness temperature difference in band E is observed for 10.5 km tangent altitude, which corresponds to an altitude inside the cloud. At 9 km tangent altitude the BT depression is larger (max. 12 K) than at 3 km tangent altitude (max. 8 K). As the propagation path through the cloud is longer for a limb scan with 9 km tangent altitude than for a scan with 3 km tangent altitude and the cloud extinction is large, more BT depression is observed at a tangent altitude of 9 km.

6.3.6 Comparison with nadir radiances

Compared to nadir radiances, limb radiances are more complex. In nadir geometry, cirrus clouds always lead to a brightness temperature depression compared to the clear sky radiances. Nadir instruments look at the tropopause and the ground, where the major source of microwave radiation is located. Clouds absorb and scatter part of the radiation out of the LOS. When we consider the lower atmosphere and the emitting ground as major sources of radiation, the propagation direction of the radiation is upwards. In nadir geometry we measure the up-welling radiation, there can not be an enhancement in this direction due to scattering according to the law of energy conservation.

In limb geometry, clouds usually lead to a brightness temperature depression if the tangent altitude lies below the cloud. But is can also lead to a BT enhancement if the tangent altitude is inside the cloud, because part of the up-welling radiation from the Earth's surface and the lower atmosphere is scattered into the LOS. In the clear sky case the sensor does not see thermal emission from the lower atmosphere at all.

The impact of cloud in nadir geometry is smaller, as the path-length through the cloud is much shorter. Of course the path-length through the cloud for a limb measurement depends on the horizontal extent of the cloud. Since clouds have an infinite extent in a 1D model, the radiances are overestimated for most cases. In reality the cloud coverage is horizontally inhomogeneous. But still the path-length through the clouds is larger in limb. The clouds presented in this study are mostly optically thin. The highest brightness temperature depression in nadir observing geometry by the clouds used to study the impact of particle size, case (1), is only 1.7 K. Only case (4), where particle size

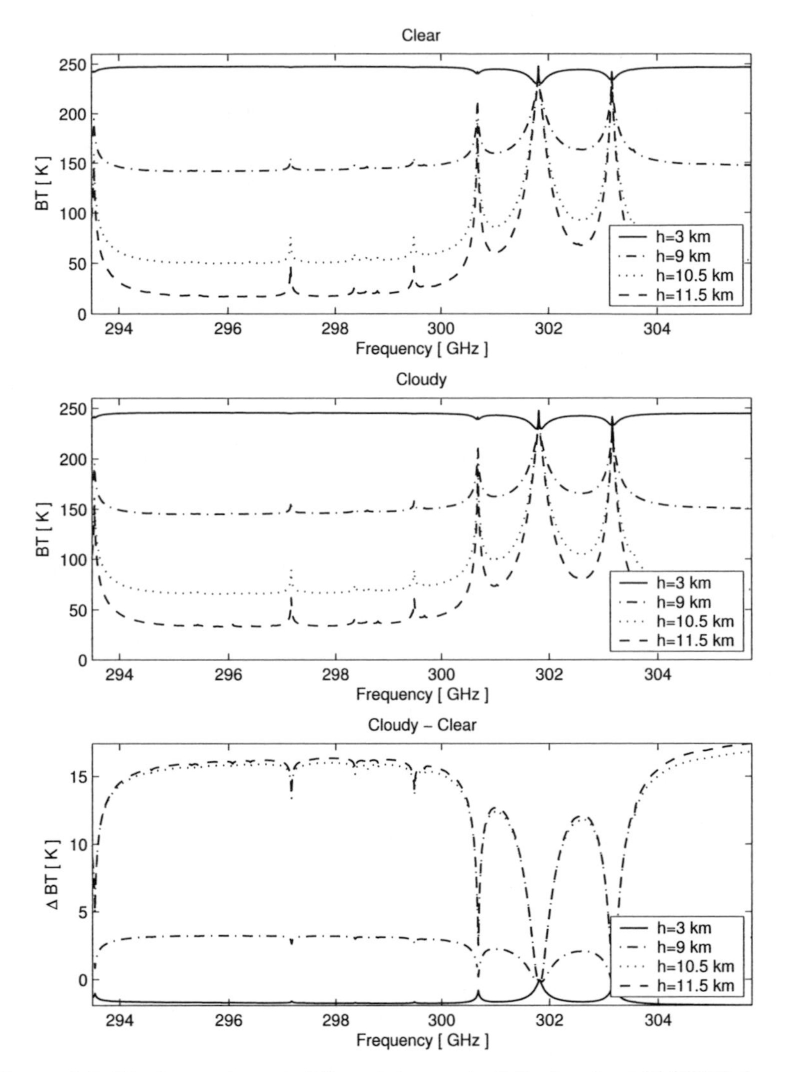

Figure 6.6: Limb spectra at different tangent altitudes for MASTER band B (294 – 304 GHz). The top panel shows the clear sky spectra, the middle panel shows the cloudy spectra and the bottom panel shows the difference between cloudy and clear sky spectra.

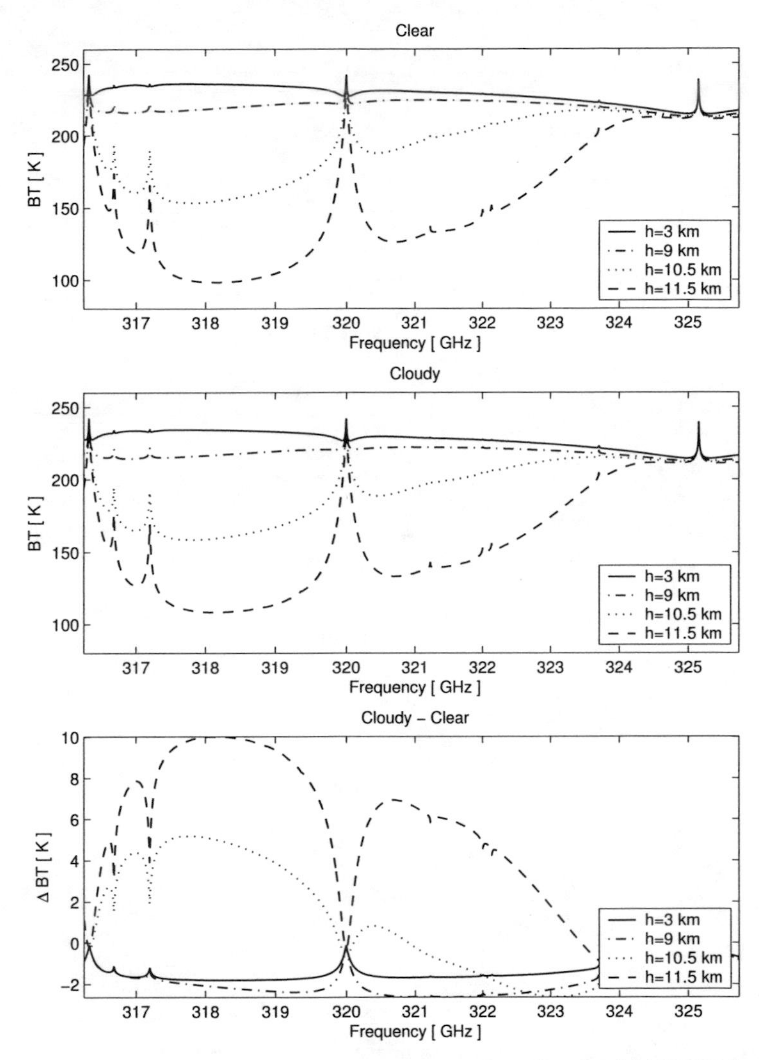

Figure 6.7: Limb spectra at different tangent altitudes. for MASTER band C (317 – 326 GHz). The top panel shows the clear sky spectra, the middle panel shows the cloudy spectra and the bottom panel shows the difference between cloudy and clear sky spectra.

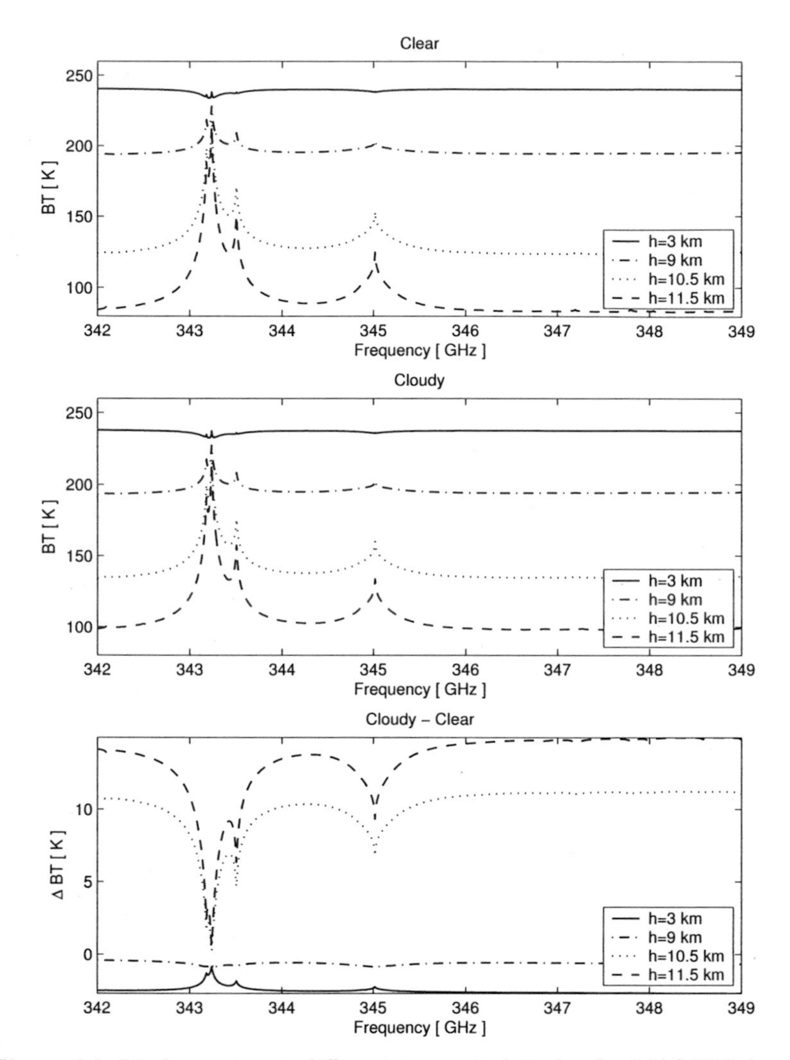

Figure 6.8: Limb spectra at different tangent altitudes for MASTER band D (342 – 349 GHz). The top panel shows the clear sky spectra, the middle panel shows the cloudy spectra and the bottom panel shows the difference between cloudy and clear sky spectra.

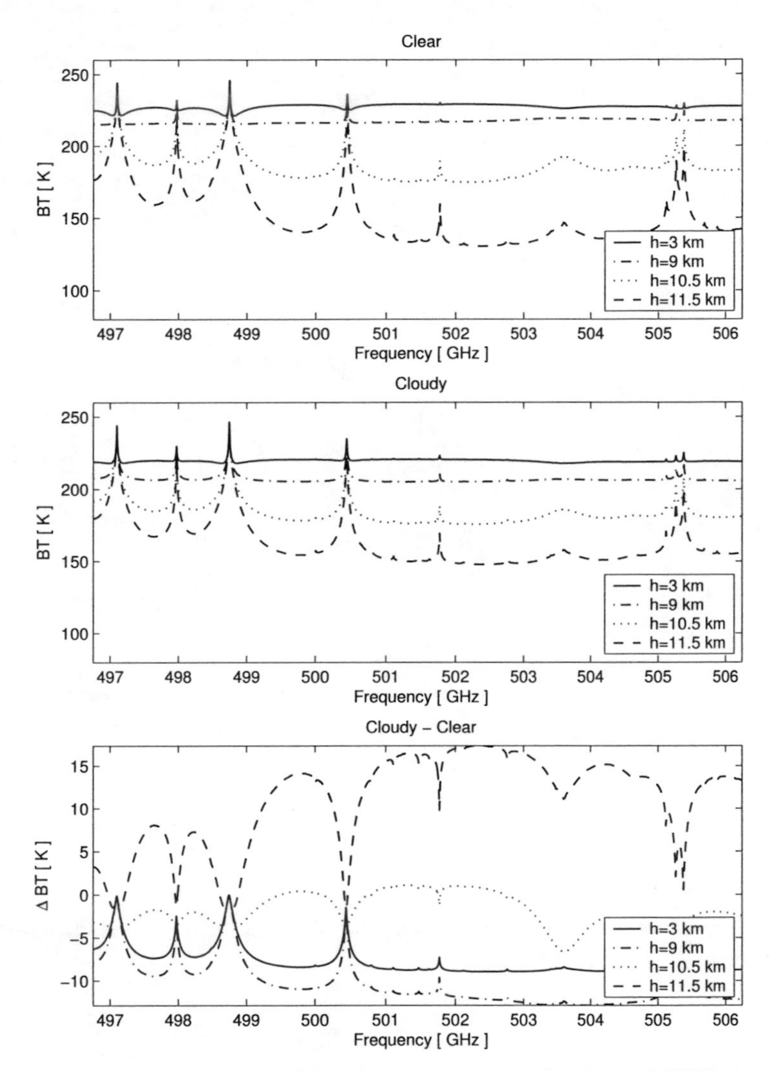

Figure 6.9: Limb spectra at different tangent altitudes. for MASTER band E (496 – 506 GHz). The top panel shows the clear sky spectra, the middle panel shows the cloudy spectra and the bottom panel shows the difference between cloudy and clear sky spectra.

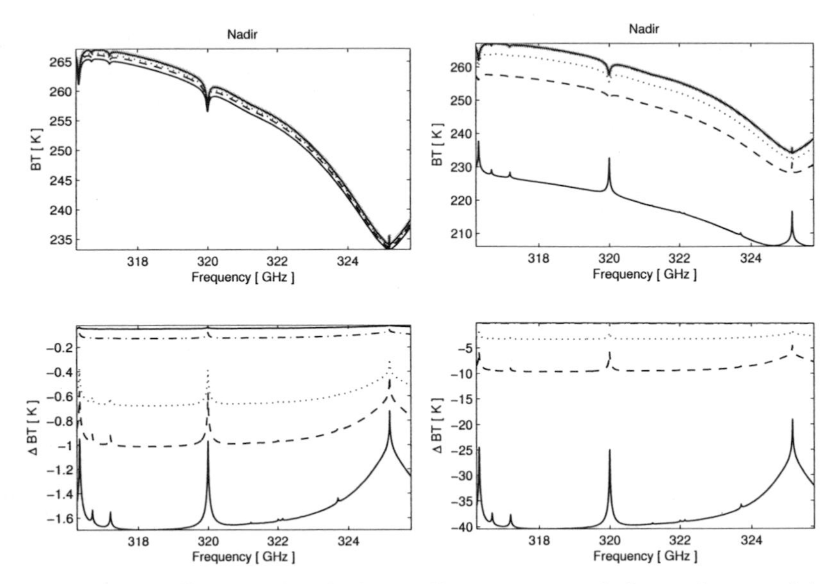

Figure 6.10: **Effect of clouds in nadir geometry.** *Left panel - case (1):* Clear sky spectrum (grey), and cloudy spectra for $R_{\text{eff}} = 21.5\,\mu\text{m}$ (—), $R_{\text{eff}} = 34.0\,\mu\text{m}$ (– · –), $R_{\text{eff}} = 68.5\,\mu\text{m}$ (· · ·), $R_{\text{eff}} = 85.5\,\mu\text{m}$ (– – –) and $R_{\text{eff}} = 128.5\,\mu\text{m}$ (—). *Right panel – case (4):* Clear sky spectrum (grey), and cloudy spectra for IMC $= 4 \cdot 10^{-5}\,\text{g/m}^3$ (—), $1.6 \cdot 10^{-3}\text{g/m}^3$ (– · –), $8 \cdot 10^{-3}\,\text{g/m}^3$ (· · ·), $0.16\,\text{g/m}^3$ (– – –) and $0.04\,\text{g/m}^3$ (—) and the corresponding particle sizes. The top plots show absolute BTs, the bottom plots show the differences between cloudy and clear sky spectra.

and IMC were increased simultaneously, shows higher BT depression (see Figure 6.10).

6.4 Discussion

The results (Figures 6.2 and 6.5) show that the ice mass content and the size of the cloud particles both are important. Consider the dotted line in Figure 6.2, corresponding to an IMC of $1.6{\cdot}10^{-3}\,\text{g/m}^3$ and an effective particle radius of $64.0\,\mu\text{m}$, and the dotted line in Figure 6.5, corresponding to the same particle size but a five times higher IMC

($8{\cdot}10^{-3}\,g/m^3$): The maximum BT depression at 8 km tangent altitude is increased by a factor of 2.5 when we compare the two results. When the effective particle size is doubled from 34.0 to 68.5 μm, the BT depression is enhanced by a factor of 9. Consequently, in these simulations the particle size is more important than the IMC. However, to estimate in detail the effect of particle size and IMC a sensitivity study for more sizes and IMC is necessary. The strong impact of both, particle size and IMC shows, that in cloud retrievals it will be a challenge to obtain IMC and particle size simultaneously, as they both lead to an enhancement of the scattering effect. For retrievals with clouds further investigations are required.

The highest effect of cloud scattering can be observed in the microwave window regions of the spectral bands. The total extinction coefficient consists of gaseous extinction and particle extinction. When the gaseous extinction cross section dominates the total extinction the scattering effect becomes smaller. In the line centers there is no difference between cloudy and clear sky spectrum. At the center frequencies the water vapor path is so high that the transmission from the cloud to the sensor is zero, thus the existence of the cloud does not affect the measured radiance. As gas absorption is very high at low altitudes the cloud altitude has a big effect on the scattering signal. At low altitudes gas absorption is dominant, therefore the scattering effect for low clouds is very small. Cirrus clouds can exist in altitudes above 10 km. Here the gas absorption is low, so the scattering signal can be very large. The scattering coefficients increase with frequency. But the scattering effect does not necessarily increase with frequency. It also depends on the gas absorption characteristics of the considered frequency region.

The particle shape is less important than the particle size and the IMC, at least for the cloud particles studied here. For higher aspect ratios and more asymmetrical particles the impact is larger, especially when the particles are oriented. For asymmetrical particles the radiation will be polarized due to scattering. First studies of the polarization characteristics are presented in the following chapters (7 and 8).

7 Simulation of polarized radiances for observations of cirrus clouds in limb- and down-looking geometry

This chapter shows first simulations using the full capabilities of the new scattering model. In the first part polarized simulations for a 1D spherical atmosphere are presented. They show the scattering and the polarization signal for limb- and down-looking geometries. Different particle sizes, shapes and orientations were considered. Furthermore the accuracy of the scalar approximation of the radiative transfer equation was validated. In the second part, polarized 3D simulations are presented. Here it was investigated in particular, how the scattering and the polarization signal depend on the sensor position with respect to the cloud. The simulations presented in this chapter have been published in Emde et al. (2004a).

7.1 Model simulations in a 1D spherical atmosphere

In all simulations it was assumed that the model atmosphere consists of nitrogen and oxygen, and the two major atmospheric trace gases: water vapor and ozone. Like in the calulations presented in the previous chapter, the concentrations are taken from FASCOD (Anderson et al. (1986)) data for mid-latitudes in summer, and gas absorption was calculated based on the HITRAN (Rothman et al. (1998)) molecular spectroscopic database using the ARTS model (version ARTS-1-0).

Atmospheric refraction was neglected. All calculations were carried out for 318 GHz. The results of the calculations are summarized in Table 7.1.

7.1.1 Scattering and polarization signal for different particle sizes

In order to study the impact of particle size on the radiation field, 1D-calculations were carried out for four different particle sizes (equal volume sphere radius): 25 µm, 50 µm, 75 µm and 100 µm. The particles are prolate spheroids with an aspect ratio of 0.5. They are either completely randomly oriented or horizontally aligned with random azimuthal orientation. The cloud altitude is 10 – 12 km and the ice mass content is $4.3 \cdot 10^{-3}$ g/m^3, which is rather small. The small value is used in order to compensate for the fact that the 1D model assumes a cloud with infinite horizontal extent. Figure 7.1 shows the radiation field just above the cloud at 13 km altitude for completely randomly oriented particles. The scattering signal increases significantly with the particle size. The top panels show the difference between the scattered intensity field and the clearsky field. At about 90° two different features can be observed: a brightness temperature (BT) enhancement or a BT depression. The physical explanation is that the main source of radiation is the thermal radiation from the lower atmosphere. For zenith angles just above 90° there is a BT enhancement because radiation coming from the lower atmosphere is scattered inside the cloud into the limb directions. This part of the radiation is missing in the down-looking directions, therefore there is a BT depression for these directions. The strongest scattering signal is observed in limb directions, since here the path-length through the cloud is the largest. The bottom panels of Figure 7.1 show the polarization signal, which is very small for randomly oriented particles. The largest polarization is observed for the largest particles ($r = 100$ µm) at about 91.5°, but even in this case it is below 1 K. The discrete jumps for zenith angles from 100° to 180° result from the polynomial interpolation of the radiation field on the cloud box boundary, which is taken as radiative background for a clear sky calculation towards the sensor. This interpolation is necessary, since the intersection zenith angle of the line of sight of the sensor with the cloud box boundary is not necessarily contained in the optimized zenith angle grid, which is used for the representation of the radiation field. Since a three point polynomial interpolation scheme is applied these jumps occur where a different set of three points is used for the interpolation. The resolution of the optimized zenith angle grid is much coarser for angles close to nadir because the radiation field does not change rapidly here. The absolute value of the jumps is very small, they can only be seen so clearly, because the scattering signal for nadir is also very small. The interpolation error is below 0.2% as shown in Figure 4.7.

Figure 7.2 shows the equivalent plots for particles, which are horizontally aligned with random azimuthal orientation. The intensity plots are similar to the cloud case with completely randomly oriented particles, but the polarization signal is much larger for oriented particles. The maximum polarization difference (Q equals the vertical minus the horizontal intensity component) is -6.3 K for the largest particles. In most regions Q is positive (partial vertical polarization), only in limb-directions just above 90° it is negative (partial horizontal polarization). For randomly oriented particles, the polarization signal is due to the radiation scattered into the line of sight, because only the phase matrix has non-zero off-diagonal elements. For horizontally aligned particles, the sign of the polarization signal is determined by two opposing mechanisms: dichroism, as manifested by a non-diagonal extinction matrix; and the effect of radiation being scattered into the line of sight. For angles just above 90° the radiation being scattered into the line of sight is the dominating mechanism, which results in a negative Q. For down-looking directions, where the cloud is between the main radiation source and the sensor the dichroism effect is dominating, which results in a positive Q. The figure shows that polarization is very significant for limb radiances when the particles are oriented.

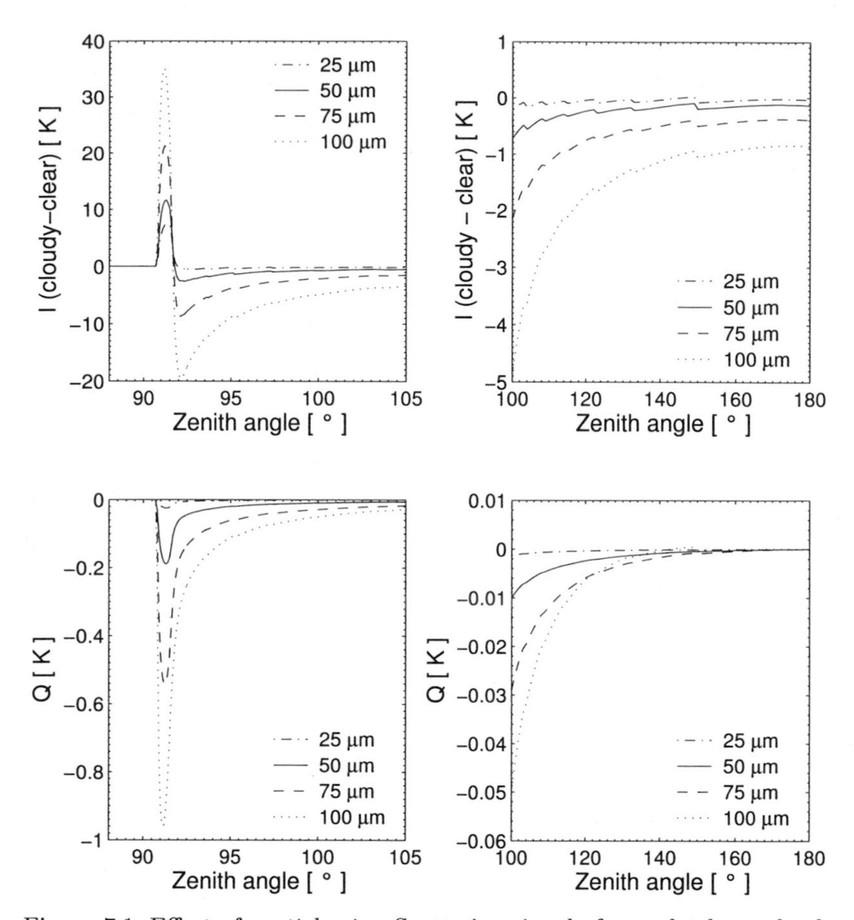

Figure 7.1: Effect of *particle size:* Scattering signal of *completely randomly oriented particles* with effective particle sizes 25 μm, 50 μm, 75 μm and 100 μm for 318 GHz at 13 km altitude. Top panels: Intensity difference between scattering calculation and clear sky calculation; bottom panels: Difference between horizontal and vertical polarization.

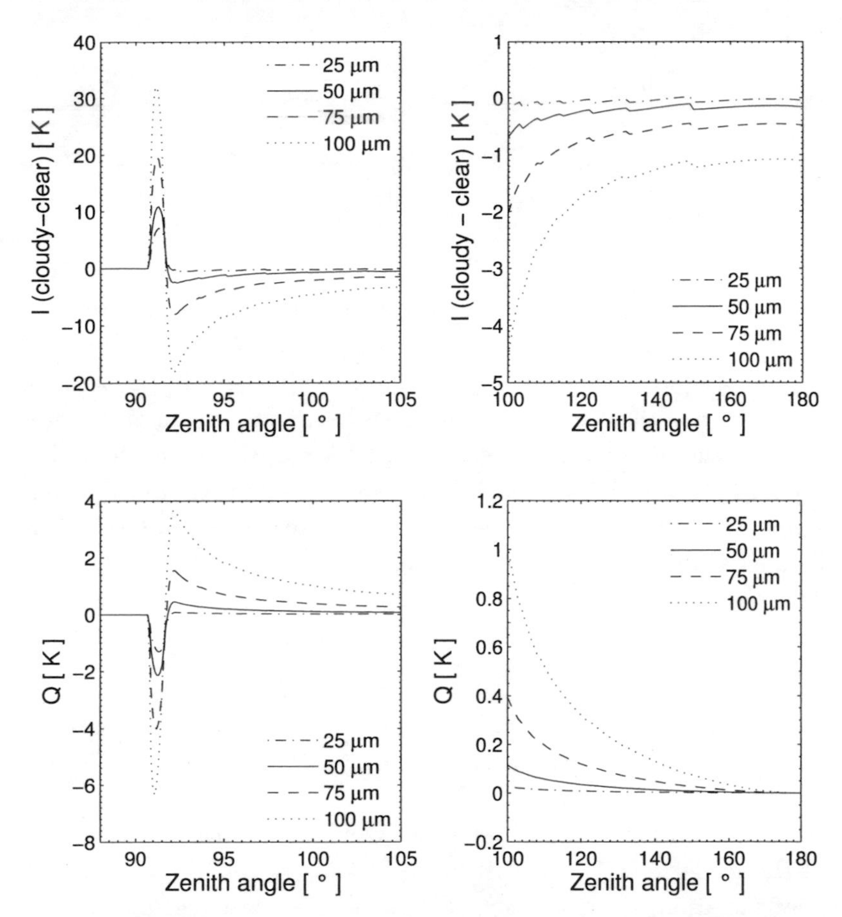

Figure 7.2: Effect of *particle size:* Scattering signal of *horizontally aligned particles* with effective particle sizes 25 μm, 50 μm, 75 μm and 100 μm for 318 GHz at 13 km altitude. Top panels: Intensity difference between scattering calculation and clear sky calculation; bottom panels: Difference between horizontal and vertical polarization.

7.1.2 Effect of particle shape

In order to look at the effect of particle shape, simulations were carried out for particles with aspect ratios 0.5 (prolate spheroids), 1.0 (spheres) and 2.0 (oblate spheroids). The particle size was 75 μm for all calculations and ice mass content and cloud height were the same as in the previous calculations. Figure 7.3 shows the results for completely randomly oriented particles. The radiation field does not change significantly for different aspect ratios. This means that the particle shape is not important for this particular setup. Figure 7.4 shows the equivalent simulations for horizontally aligned particles with random azimuthal orientation. Here there are significant differences between the different particle shapes. The intensity plots show that the BT enhancement and the BT depression are similar for all particle shapes (cf. Table 7.1). The maximum absolute values of Q are -6.4 K and -4.0 K for oblate and prolate spheroids respectively. For spherical particles there is only a very small polarization signal. More simulations are required to study in detail the effect of particle shape on the polarization signal.

7.1.3 Scalar simulations

In order to save CPU time and memory one can use the scalar version of the model (cf. Section 4.1.3). To test the accuracy of the scalar approximation, all calculations presented above were performed using the scalar version. Figure 7.5 shows the differences between the scalar and the vector calculations for completely randomly oriented particles with different effective radii. The maximum difference for limb directions is 0.01 K and for down-looking directions $7{\cdot}10^{-4}$ K. These small differences show, that it is not necessary to use the fully polarized vector version to model the radiative transfer through scattering media with completely randomly oriented particles. The previous section has also shown, that the polarization signal is negligible for such cases. Figure 7.6 shows the equivalent results for horizontally aligned particles. For down-looking directions the difference is below 0.02 K, but for limb-cases it can go up to 1.5 K. For this reason one should use

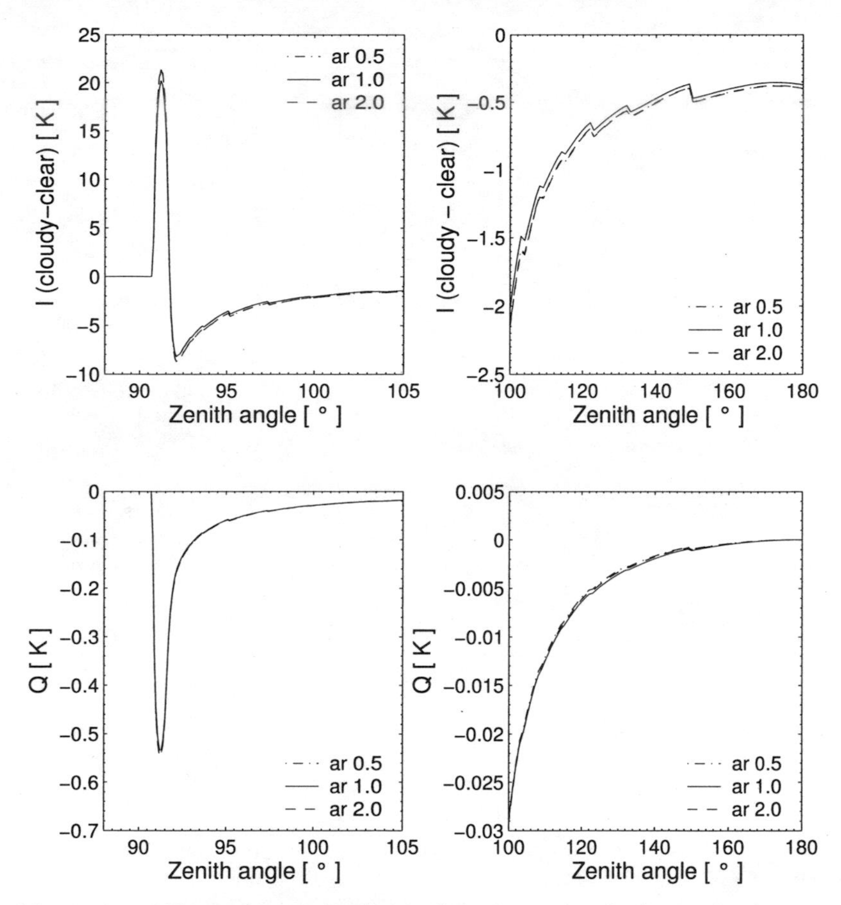

Figure 7.3: Effect of *particle shape:* Scattering signal of *completely randomly oriented spheroidal particles* with aspect ratios 0.5, 1.0 and 2.0 for 318 GHz at 13 km altitude. Top panels: Intensity difference between scattering calculation and clear sky calculation; bottom panels: Difference between horizontal and vertical polarization.

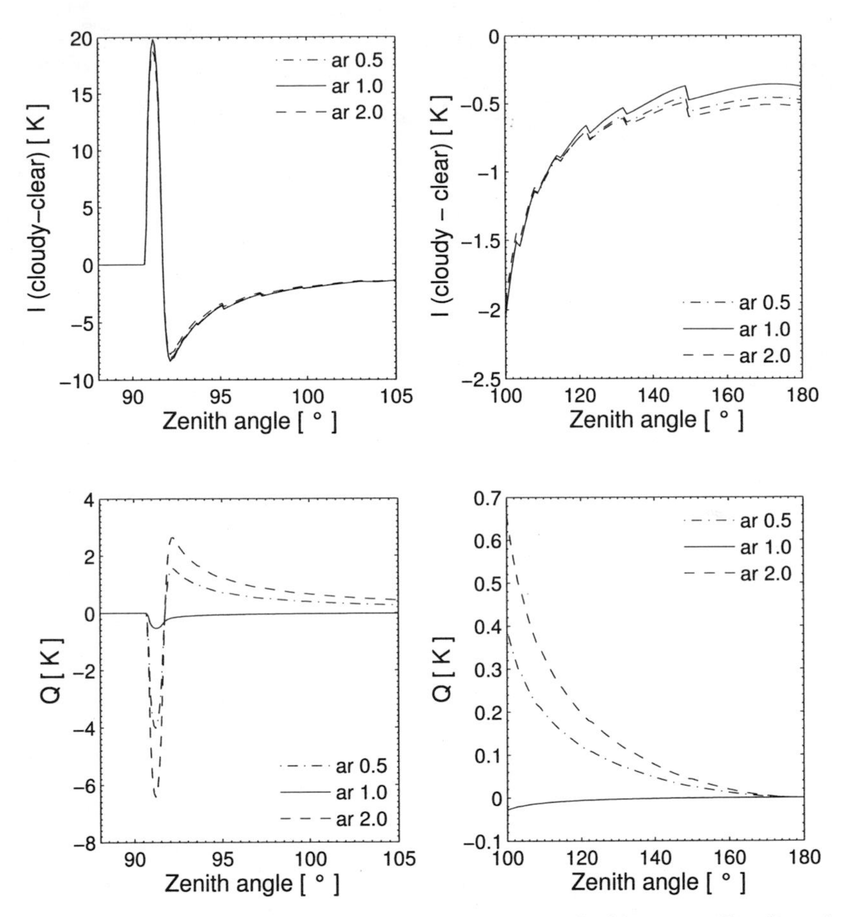

Figure 7.4: Effect of *particle shape:* Scattering signal of *horizontally aligned spheroidal particles* with aspect ratios 0.5, 1.0 and 2.0 for 318 GHz at 13 km altitude. Top panels: Intensity difference between scattering calculation and clear sky calculation; bottom panels: Difference between horizontal and vertical polarization.

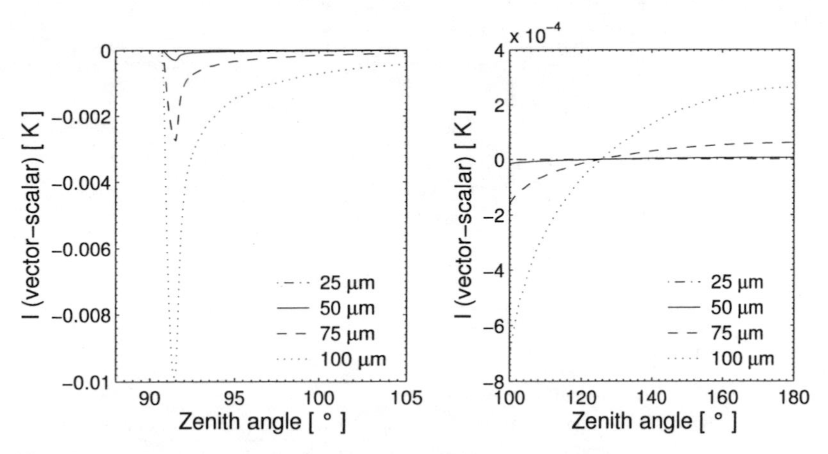

Figure 7.5: Difference between vector RT and scalar RT calculations for completely randomly oriented spheroidal particles (aspect ratio 2.0) for 318 GHz at 13 km altitude.

the vector model for limb RT simulations through scattering media consisting of oriented particles even if one is only interested in the total intensity of the radiation.

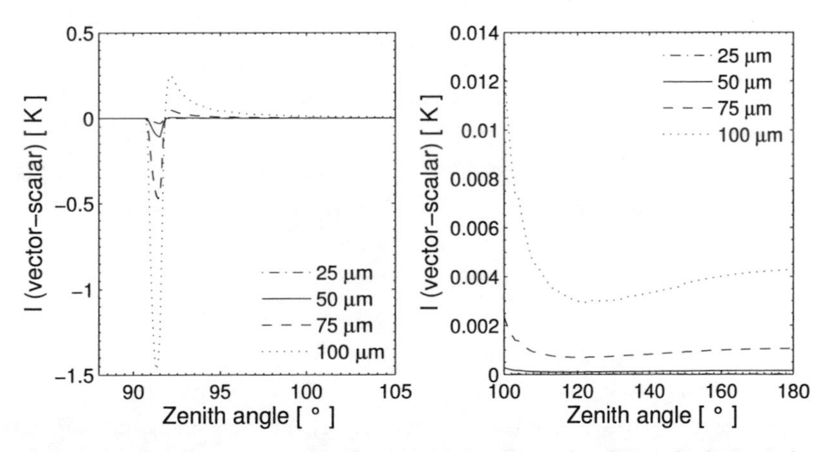

Figure 7.6: Difference between vector RT and scalar RT calculations for horizontally aligned spheroidal particles (aspect ratio 2.0) for 318 GHz at 13 km altitude.

Table 7.1: Summary of simulations

Size	Aspect	BT enh.	BT dep.		Polarization	
[μm]	ratio	ΔBT_{max}	ΔBT_{max}	$\Delta BT_{120°}$	Q_{max}	$Q_{120°}$
	[−]	[K]	[K]	[K]	[K]	[K]
p20: Completely randomly oriented particles						
25		7.56	-0.49	-0.04	-0.03	-0.00
50	0.5	11.67	-2.61	-0.23	-0.19	-0.00
75		21.32	-8.72	-0.75	-0.54	-0.01
100		35.04	-19.63	-1.72	-0.96	-0.01
	0.5	21.32	-8.72	-0.75	-0.54	-0.01
75	1.0	20.18	-8.21	-0.70	-0.53	-0.01
	2.0	21.45	-8.78	-0.76	-0.55	-0.01
p30: Horizontally aligned particles with random azimuthal orientation						
25		7.11	-0.47	-0.04	-1.31	0.01
50	0.5	10.88	-2.47	-0.23	-2.13	0.03
75		19.65	-8.18	-0.75	-4.01	0.12
100		32.03	-18.29	-1.73	-6.33	0.32
	0.5	19.65	-8.18	-0.75	-4.01	0.12
75	1.0	19.82	-8.36	-0.71	-0.52	-0.01
	2.0	18.79	-7.78	-0.75	-6.42	0.20

7.2 3D box type cloud model simulations

The 3D version of the model was applied for simulating limb radiances for a cloud of finite extent embedded in a horizontally homogeneous atmosphere. The height of the cloud box was 7.3 to 12.5 km and the vertical extent of the cloud was from 9.4 to 11.5 km. The latitude range was 0° to 0.576° and the longitude range was 0° to 0.288°. This corresponds to a horizontal extent of approximately 64 km × 32 km. A coarse spatial discretization was chosen, because a fine resolution is not necessary when the cloud is homogeneous; the number of grid points was 6 × 9 × 5. Simulations were performed for two different IMC: 0.02 g/m^3 and 0.1 g/m^3 corresponding to limb optical depths of approximately 0.5 and 2.8 respectively. The maximum propagation path step length was set to 1 km for the optically thin cloud and to 250 m for the optically thicker cloud. These values allow to assume single scattering for each propagation path step. It was assumed that the cloud consists of spheroidal ice particles with a particle size of

75 μm and an aspect ratio of 0.5. Calculations were performed for completely randomly oriented particles and for horizontally aligned particles with azimuthally random orientation. The sensor was placed on board a satellite following a polar orbit at 820 km altitude. At each sensor position tangent altitudes from 0 to 13 km were measured. Figure 7.7 shows corresponding lines of sight (LOS). The figure shows that the cloud is seen from different sides, from the top, from the bottom or from the left side. When the satellite is at a latitude of 25° the cloud is only seen for low tangent altitudes (from 0 to 6 km). The cloud is seen at higher tangent altitudes at 27.5°. For even greater latitudes the sensor sees the cloud from the bottom. Note that the tangent point in the first plot is behind the cloud, in the second plot in the middle of the cloud and in the last plot in front of the cloud. In order to compare the 1D and the 3D model versions, simulations for a 1D cloud layer with equivalent limb optical depths at 10 km tangent height were performed. The IMC for the equivalent 1D clouds were $0.005\,\mathrm{g/m^3}$ and $0.025\,\mathrm{g/m^3}$.

Figures 7.8 to 7.10 show the simulated radiances plotted as a function of tangent altitude and sensor position for totally randomly oriented particles. The top panels show the intensity differences between the clear sky calculation and the cloudy sky calculation. The bottom panels show the polarization difference Q. The contour plots on the left hand side are the 3D results. Reddish colors indicate a brightness temperature enhancement due to the cloud and bluish colors indicate a BT depression. White means that there is no cloud effect. A cloud effect can only be seen at tangent heights for which the corresponding lines of sight intersect the cloud. The intensity plot shows that up to a latitude of 27° there is a BT depression due to the cloud. The reason is that in those cases the tangent point, from where the major source of thermal radiation emerges, is behind the cloud. The cloud scatters part of the radiation away from the line of sight. For latitudes above 28° a BT enhancement is observed. In these cases the tangent point is in front of the cloud. The sensor measures all radiation emerging from the tangent point and additionally the back-scattered radiation from the cloud behind the tangent point. If the tangent point is inside the cloud, between 26.5° and 28°, a BT enhancement can be observed for high tangent altitudes because part of the up-welling radiation from the lower atmosphere is scattered into the direction of the LOS. For lower tangent points the scattering away from the LOS dominates, hence a BT depression is observed in this latitude range. The maximum absolute values for the BT enhancement and the BT depression are 19 K and −23 K respectively. The equivalent 1D result on the right hand side shows a larger BT enhancement of 22 K and a smaller BT depression of −10 K. The BT depression is smaller because the optical depth for tangent heights below the cloud is smaller in the 1D calculation compared to the 3D calculation with much larger IMC. The polarization plots shows that in the 3D case as well as in the 1D case there is only a very small polarization difference for totally randomly oriented particles. In 3D, it is between −0.4 K and 0.1 K and in 1D between −0.5 K and 0 K. In 1D, only negative polarization is observed whereas in 3D it can be positive or negative. The intensity plot in Figure 7.9 for azimuthally randomly oriented particles looks similar to that for completely randomly oriented particles. However, the maximum values of BT enhancement and BT depression are slightly smaller, about 17 K and −22 K respectively. In 1D, the intensity differences are in the range of −9 K to 20 K. The polarization difference becomes much larger, between −3.5 K and 4.0 K can be observed in the 3D simulation and between −4.0 K and 1.7 K in the equivalent 1D simulation.

Figure 7.10 shows the results of the simulation for the thicker cloud consisting of horizontally aligned particles. The pattern looks very similar to that obtained for the thinner cloud but the absolute values of the BT depression, the BT enhancement and the polarization are much larger. The intensity difference is in the range from -63 K to 45 K and the polarization difference is in the range from −7.0 K to 5.2 K for the 3D calculation. The equivalent 1D result ranges from −35 K to 55 K for the intensity and from −7.0 K to 5.2 K for the polarization difference. Since the pattern for the thicker cloud is similar to that obtained in the thin cloud case also for randomly oriented particles, the plot is not inlcuded here. The intensity difference ranges in this case from −65 K to 47 K for 3D and from −37 K to 58 K for 1D. The polarization difference ranges from $-0.7\,\mathrm{K}$ to $0.8\,\mathrm{K}$ for 3D and from $-1.0\,\mathrm{K}$ to $0\,\mathrm{K}$ for 1D.

Overall the comparison between 1D and 3D shows similar results at tangent heights inside the cloud, where the optical depth is approximately equivalent. For other tangent heights, the optical depths are different and therefore the results deviate strongly. The scattering signal in 3D depends very much on the sensor position w.r.t. the cloud. Hence it is very important to use the 3D model where the cloud extent is not very large, like in this example case, or where the clouds are horizontally inhomogeneous.

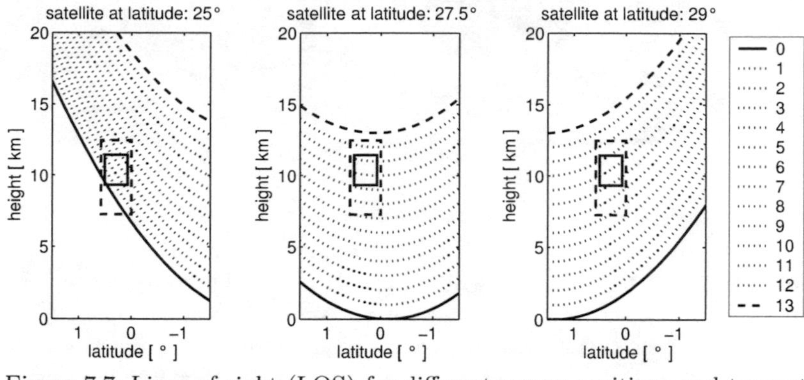

Figure 7.7: Lines of sight (LOS) for different sensor positions and tangent altitudes [km]. The solid line corresponds to a LOS for a tangent altitude of 0 km and the dashed line to a LOS for a tangent altitude of 13 km. Dotted lines correspond to LOS for tangent heights between 0 and 13 km. Inside the solid rectangle the single scattering properties are defined and the dashed rectangle labels the cloud box. Courtesy of Claas Teichmann.

7.2.1 Performance

The CPU time for the thin cloud cases was approximately 50 minutes on a 3 GHz Pentium 4 processor, when all four Stokes components were calculated. U and V are not discussed as they are approximately zero (less than $10^{-7}\,\mathrm{K}$) for all calculations. The computation time can be reduced by 25% without loosing accuracy when one runs the model

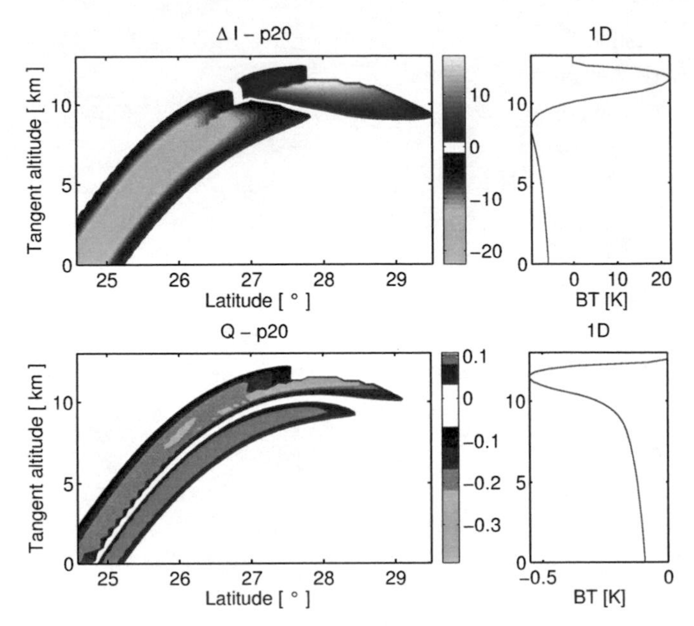

Figure 7.8: Left panels: Scattering signal of a 3D box-type cloud embedded in a 1D atmosphere as a function of sensor position and tangent altitude for 318 GHz. The cloud consists of completely randomly oriented spheroidal particles with a size of 75 μm and with an aspect ratio of 0.5. The IMC is $0.02\,\mathrm{g/m^3}$. Right panels: 1D result for a cloud with an equivalent optical depth in limb (IMC $= 0.005$ g/m^3). The upper plots show the intensity I and the lower plots the polarization difference Q.

only for two Stokes components. The computation time for the same scenario was in this case approximately 37 minutes. The calculation for the thicker cloud took much longer, approximately 150 minutes for all four Stokes components, because the maximum propagation-path step length needed to be reduced.

The computation time increases strongly with the size of the cloud box. Doubling the number of grid points in one dimension means a doubling of the computation time. Therefore the 3D version of the model can be used for accurate simulations to study the effect of cloud inhomogeneity, but it is not applicable for operational use. The performance of the 1D version of the model is much better. All 1D

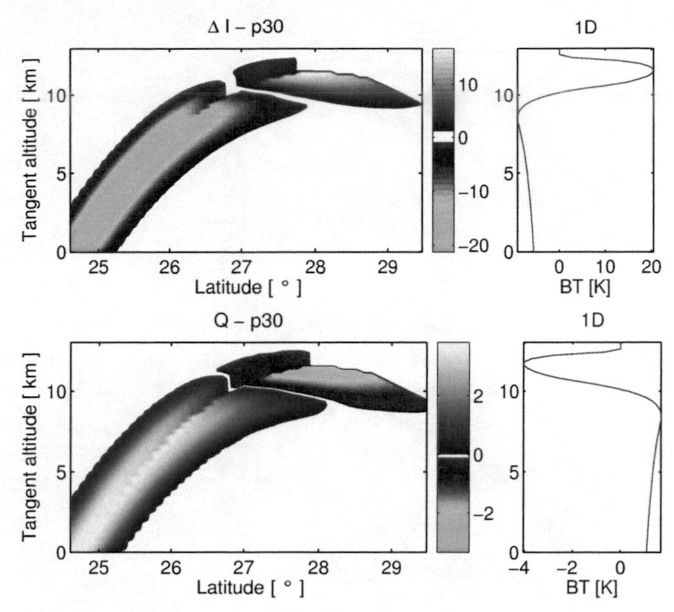

Figure 7.9: Left panels: Scattering signal of a 3D box-type cloud embedded in a 1D atmosphere as a function of sensor position and tangent altitude for 318 GHz. The cloud consists of horizontally aligned spheroidal particles with a size of 75 μm and with an aspect ratio of 0.5. The IMC is 0.02 g/m^3. Right panels: 1D result for a cloud with an equivalent optical depth in limb (IMC = 0.005 g/m^3). The upper plots show the intensity I and the lower plots the polarization difference Q.

simulations shown in this chapter needed less than 30 seconds CPU time.

7.3 Conclusions

For the unpolarized simulations being presented in Chapter 6 as well as for the polarized simulations presented in this chapter, the particle size strongly influences the scattering signal. The polarization signal also depends strongly on the particle size. Particle shape is an important cloud parameter when the cloud particles are horizontally aligned with random azimuthal orientation. In the case of totally randomly

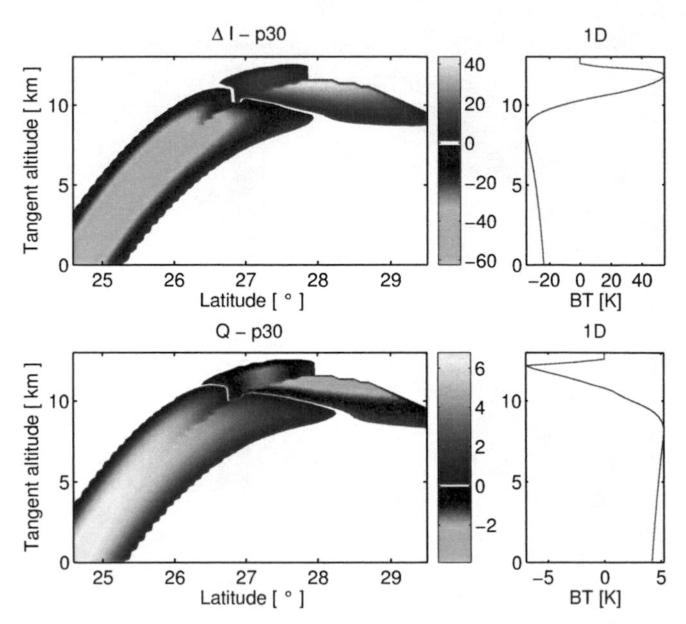

Figure 7.10: Left panels: Scattering signal of a 3D box-type cloud embedded in a 1D atmosphere as a function of sensor position and tangent altitude for 318 GHz. The cloud consists of horizontally aligned spheroidal particles with a size of 75 μm and with an aspect ratio of 0.5. The IMC is 0.1 g/m^3. Right panels: 1D result for a cloud with an equivalent optical depth in limb (IMC = 0.025 g/m^3). The upper plots show the intensity I and the lower plots the polarization difference Q.

oriented particles, changing the particle shape shows almost no effect in the simulations. For horizontally aligned particles, there is a significant difference between the scalar (unpolarized) version and the vector (polarized) version of the model in intensity. Therefore it is important to use a vector radiative transfer model to obtain accurate results, even if one is only interested in intensity, not in polarization. The 3D simulations show that one must not neglect cloud inhomogeneity effects. The scattering signal depends strongly upon the sensor position with respect to the cloud. The fact that the scattering signal is much larger in limb geometry compared to down-looking geometries, due to the greater path-length through the cloud layers, demonstrates the potential of retrieving cloud properties from limb measurements. The major disadvantage of the 3D DOIT model is that is not yet useful for operational applications due to the large computation time. But it is practical for research, for instance to study in detail the effect of different cloud parameters on polarization. The performance of the 1D model is much better than that of the 3D model. Therefore the 1D version of the model can be applied to calculate full frequency spectra or for detailed cloud sensitivity studies.

8 A study to investigate the impact of thin layer cirrus clouds in tropical regions on the EOS MLS instrument

The Microwave Limb Sounder (MLS) experiments perform measurements of atmospheric composition, temperature, and pressure by limb observations of millimeter- and sub-millimeter-wavelength thermal emission. The first MLS experiment in space was launched on the Upper Atmospheric Research Satellite (UARS) in 1991. A follow-on MLS instrument was developed for NASA's Earth Observing System (EOS). The EOS MLS instrument was launched on the EOS Aura satellite on the 15th of July, 2004. EOS MLS is a passive instrument that has radiometers in spectral bands centered near 118, 190, 240, 640 and 2500 GHz. Information about the UARS and the EOS MLS instruments is given in Waters et al. (1999). This chapter presents scattering simulations for the EOS MLS instrument with focus on sensitivity of the scattering and polarization signal on ice mass content and aspect ratio. Furthermore the possibility of retrieving shape information by combination of channels with different polarization characteristics is investigated.

8.1 Setup

8.1.1 Selection of frequencies

The EOS MLS spectral channels, which were selected for the study, are given in Table 8.1. R1A and R1B both measure at 122 GHz, where

R1A measures the vertically polarized component of the intensity and R1B the horizontally polarized component. Hence the combination of the measurements of the two radiometers gives the polarization difference Q at 122 GHz. Furthermore we selected the 200.5 GHz channel of R2, which measures the vertically polarized part, and the 230 GHz channel of R3, which measures the horizontally polarized part of the radiation.

Radiometer (Polarization)	Frequency	Required molecules
R1A (V)	122 GHz	O_2, N_2, H_2O
R1B (H)	122 GHz	same as R1A
R2 (V)	200.5 GHz	O_2, N_2, H_2O
R3 (H)	230 GHz	O_2, N_2, H_2O, CO, O_3

Table 8.1: Selected EOS MLS spectral channels for sensitivity study

8.1.2 Particle size distribution function

For all simulations the particle size distribution by Mc Farquhar and Heymsfield, which was introduced in Section 3.5.3, was used. For each simulation one particular particle shape was assumed. In order to study the effect of particle shape, many simulations were performed using the size distribution by Mc Farquhar and Heymsfield for aspherical particles with different aspect ratios. Moreover it was assumed that the particles are horizontally aligned.

8.1.3 Included species for absorption coefficient calculations

Gaseous species, which have important absorption lines or continua at the frequencies of the considered channels, are listed in Table 8.1. All of these species were included in the simulations. The selection of gaseous species is based on Waters et al. (1999). Atmospheric profiles were taken from the FASCOD (Anderson et al., 1986) data for tropical regions. The water vapor profile was adjusted so that the relative humidity was 100% with respect to ice at altitudes with non-zero ice mass content.

8.1.4 Definition of realistic cloud parameters

For this study box shaped clouds in pressure, latitude and longitude are considered. The question of cloud inhomogeneity is neglected in order to simplify the interpretation of the results with respect to the effect of IMC, particle shape and the cloud position relative to the sensor. Thin layer cirrus clouds are rather homogeneous and have a large horizontal extent, hence the homogeneous box cloud is a good approximation for such kind of cloud. It might even be possible to use the 1D model, which is much faster than the 3D model. From observations, Del Genio et al. (2002) have derived typical cirrus cloud altitude and IMC ranges. These have been used for all simulations. The altitude range was set from 11.9 to 13.4 km. The IMC ranged from 0.0001 to 1 g/m^3. In order to study the effect of IMC several simulations for homogeneous clouds with different IMC were performed. The horizontal extent in all 3D simulations was 400 km $\times$ 400 km, which corresponds to 3.6° latitude times 3.6° longitude in the tropics.

8.2 Clear sky and cloudy radiances at 122, 200.5 and 230 GHz

As a first step the clear sky and cloudy radiances for the three channels (122, 200.5 and 230 GHz) were calculated for one special cloud case using the 1D model in order to see qualitatively the different behavior of the channels. The IMC was 0.1 g/m^3 and the aspect ratio was 1.5. The results are shown in Figure 8.1. The top left panel shows the clear sky radiances for tangent altitudes from 1 to 13 km. The figure shows, that for 122 GHz the gas absorption at high tangent altitudes is larger compared to 200.5 and 230 GHz so that saturation is reached at a higher altitude. The clear sky radiances for tangent altitudes higher than approximately 8 km are larger for 122 GHz than for 200.5 and

230 GHz, and for tangent altitudes below approximately 8 km the opposite is observed. The radiances obtained for 200.5 and 230 GHz are very similar to each other, that means that these two channels have similar absorption features. The top right panel shows cloudy radiances for the three channels and the bottom left panel shows the difference between cloudy and clear sky radiances. Obviously at 122 GHz the cloud has much less effect compared to the other channels. The BT depression at low tangent altitudes is larger for 230 GHz compared to 200.5 GHz, which is mostly due to an increasing extinction coefficient. The bottom right panel shows the polarization difference Q. Again this is largest for 230 GHz. Also here 200.5 and 230 GHz behave similarly and 122 GHz looks completely different and shows a much smaller polarization signal. A positive polarization difference is due to extinction by scattering, the radiation scattered away from the propagation direction is horizontally polarized, which means that vertically polarized radiation is left in the propagation direction. The radiation scattered into the propagation direction is horizontally polarized. Therefore, the polarization difference $Q = I_v - I_h$ is positive, if more radiation is scattered away from the line of sight (LOS) than into the LOS and negative if more radiation is scattered into the LOS.

8.3 Comparison between 1D and 3D simulations

In order to find out, whether it is appropriate to use the 1D model for the thin cirrus cloud layer, calculations were performed for different sensor positions with respect to the cloud. The LOS of the sensor at different positions intersect the cloud box at different points illustrated in Figure 8.2. LOS sets A and B were also used for the comparison of the DOIT module with the Monte Carlo module, which was presented in Section 5.3. The bottom panel of Figure 5.8 shows the cloud box and the LOS sets A and B. Intersection point A is 50 km away from the north edge of the cloud, B is in the middle of the cloud, C is 100 km away from the south edge of the cloud and D is 50 km away from the

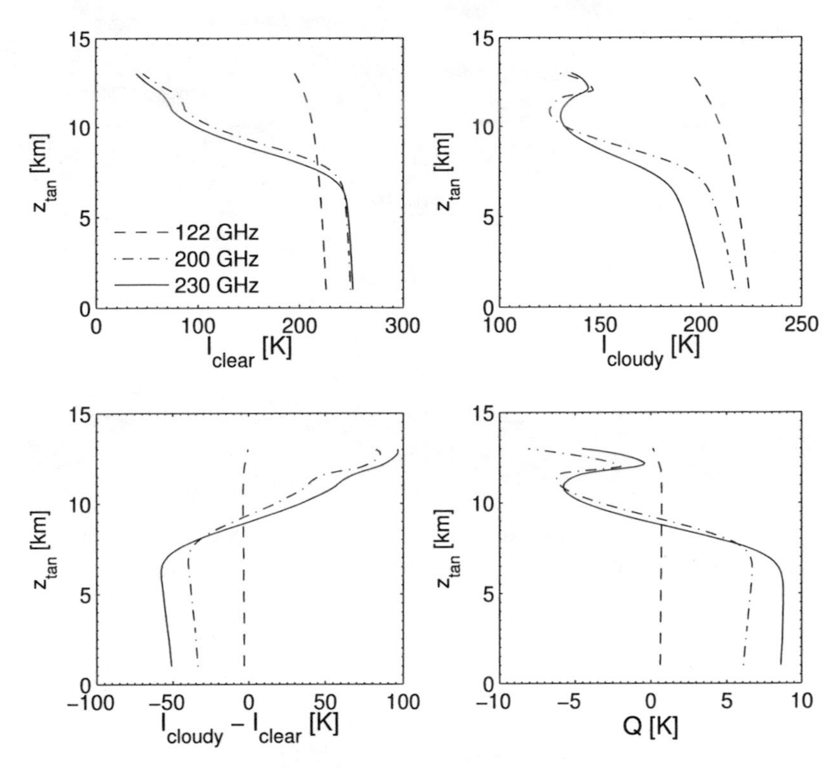

Figure 8.1: Top left: Clear sky radiances. Top right: Cloudy radiances. Bottom left: Cloudy minus clear sky radiances. Bottom right: Polarization difference.

east edge of the cloud. A single aspect ratio of 1.5 was applied for the whole cloud. The ice mass content was assumed to be $0.1\,\mathrm{g/m^3}$.

The differences between the 1D and the 3D DOIT model for all LOS sets are shown in Figure 8.3. For 122 GHz the difference is for the intensity I less than 0.2 K and for the polarization signal Q less than 0.1 K.

For 230 GHz and LOS set C the difference for I is more than 10 K and for Q it is more than 4 K. For the other LOS sets the differences are much smaller, but still significant, especially at 12 km tangent altitude. At 13 km tangent altitude the largest difference is obtained for LOS set A and at 12 km tangent altitude the largest difference is obtained

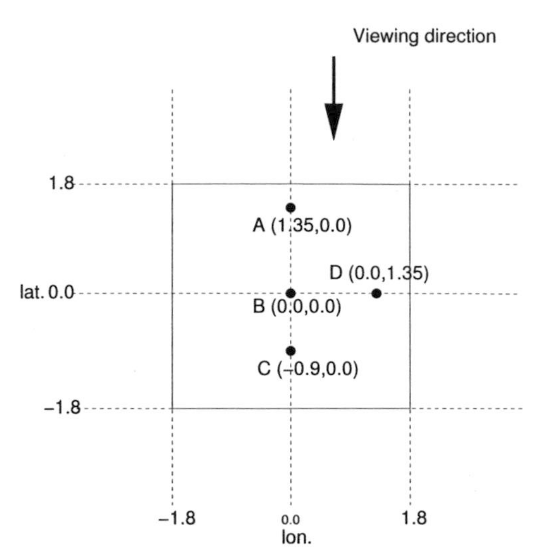

Figure 8.2: Cloud box and crossing points of LOS sets A, B, C and D as seen from the top. The arrow shows the viewing direction of the instrument.

for LOS set C. This can be explained by looking at the LOS relative to the cloud box. For A the LOS corresponding to 13 km tangent altitude intersects the cloud latitude boundary, hence the path-length through the cloud is shorter in the 3D model. Therefore, at 122 GHz, where the cloud leads to BT depression, the 1D model yields smaller BT compared to the 3D model. On the contrary, at 230 GHz, where the cloud leads to a BT enhancement, the 1D model yields larger BT. The same is valid for 12 km tangent altitude of LOS set C. Altogether, LOS set C shows larger differences than LOS set A, although the intersection point is further away from the cloud box boundary.

Apart from LOS set C the differences between the 1D and the 3D model are smaller than the differences between the 3D model and the 3D Monte Carlo model. Therefore, numerical inaccuracies are larger than the error introduced by the 1D approximation. This shows, that it is reasonable to use the 1D model for studying the effect of cloud parameters of the thin cirrus layer.

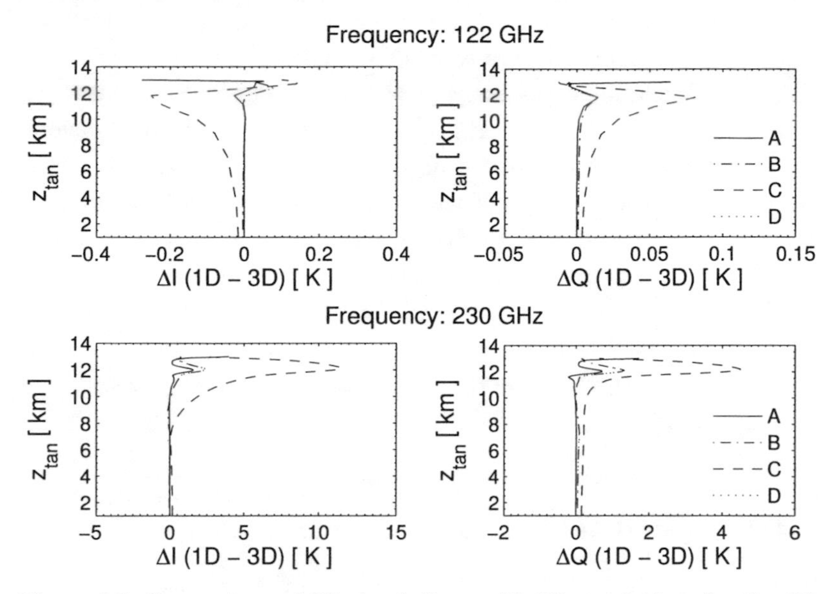

Figure 8.3: Comparison of 3D simulations with 1D simulations for the different sets of LOS. The top panels show the results for 122 GHz and the bottom panels show the results for 230 GHz.

8.4 Sensitivity study

8.4.1 Dependence on ice mass content

In order to study the dependence of the total intensity and the polarization difference Q on the IMC, we plotted simulations for a constant aspect ratio of 3.0 and varied the IMC from 0.01 to 1 g/m^3. The results for all channels are presented in Figure 8.4. The left panels show the difference between the cloudy and the clear sky intensities and the right panels show the polarization differences. For 122 GHz the BT depression due to the cloud increases monotonically with increasing IMC. The polarization difference is positive and also increases monotonically with IMC. The middle and bottom panels show the results for 200.5 and 230 GHz respectively. The patterns look very similar to each other but different to the 122 GHz case. For intensities, as we have already seen in Figure 8.1, there is a BT enhancement at high tangent altitudes and a BT depression at low tangent altitudes. The BT depression increases with IMC. The enhancement at 12 km tangent altitude increases up to 0.2 g/m^3, but for even higher IMC it decreases again. The reason is that the optical depth of the cloud is so large that multiple scattering events from the lower part of the cloud do not reach the top of the cloud. The polarization signal for small tangent altitudes is positive and increases up to approximately 0.2 g/m^3, then it starts decreasing. Due to multiple scattering events the polarization signal is decreased. For high tangent altitudes the polarization signal changes the sign from negative to positive at approximately 0.2 g/m^3. The optical depth of the cloud increases, so that less radiation is scattered into the LOS.

8.4.2 Dependence on aspect ratio

In order to study the effect of aspect ratio on the polarized radiances, simulations for an IMC of 0.1 g/m^3 are presented in Figure 8.5. The aspect ratio is varied from 1/5 to 5. The differences between cloudy and clear sky radiances on the left hand side show that the effect of particle shape is very small. To study the impact of particle shape the polarization signal is much more interesting. At all frequencies, for particles with an aspect ratio of one the polarization signal is negligible and it increases in both deformation directions. At 122 GHz the polarization signal is always positive. It increases up to approximately 2 K for oblate spheroids and up to approximately 1.5 K for prolate spheroids. The same behavior is seen for tangent altitudes below 8 km for the frequencies 200.5 and 230 GHz. However, the signal is much larger, up to approximately 20 K for 200.5 GHz and up to 25 K for 230 GHz. The absolute value of the negative polarization difference at high tangent altitudes is smaller for 12 km tangent altitude than below and above this altitude, because the path-length through the cloud is the largest in this case, which means that much radiation is multiple scattered. This decreases the polarization signal. Again the polarization signal is larger for oblate than for prolate particles with the same deformation.

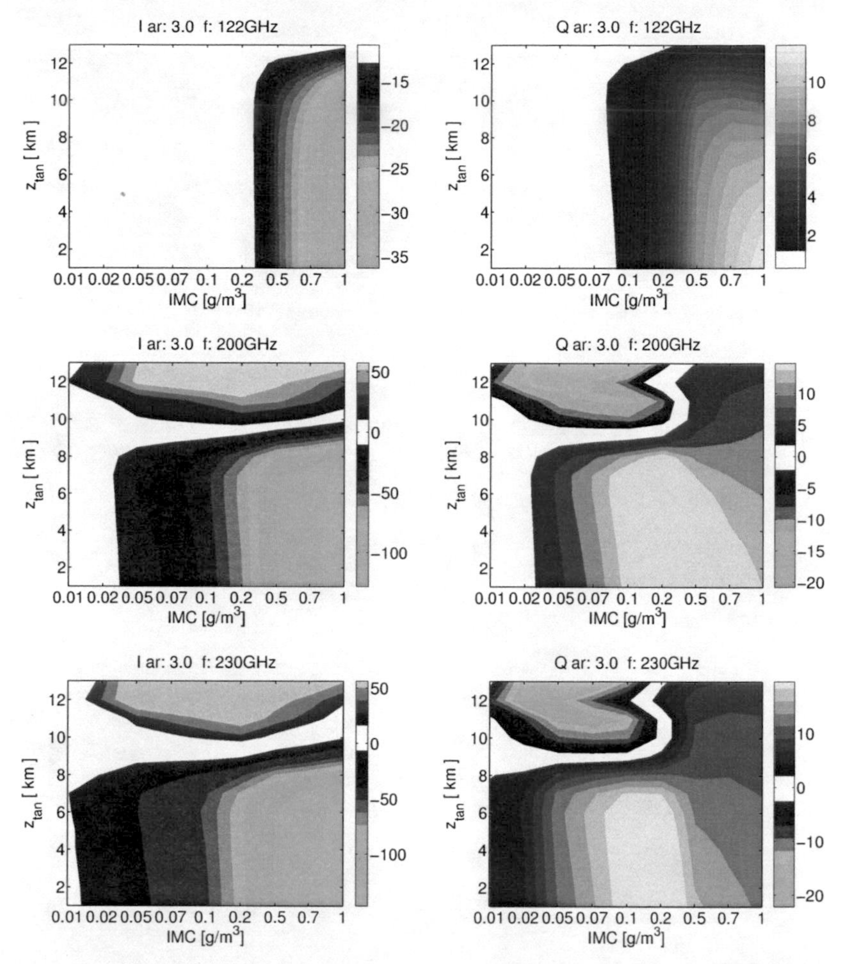

Figure 8.4: Dependence of total intensity and polarization on IMC at 122 GHz, 200.5 GHz, and 230 GHz for a cloud consisting of plates with an aspect ratio of 3.0. The contours correspond to the intensity difference compared to clear sky radiances (left) and to the polarization difference (right).

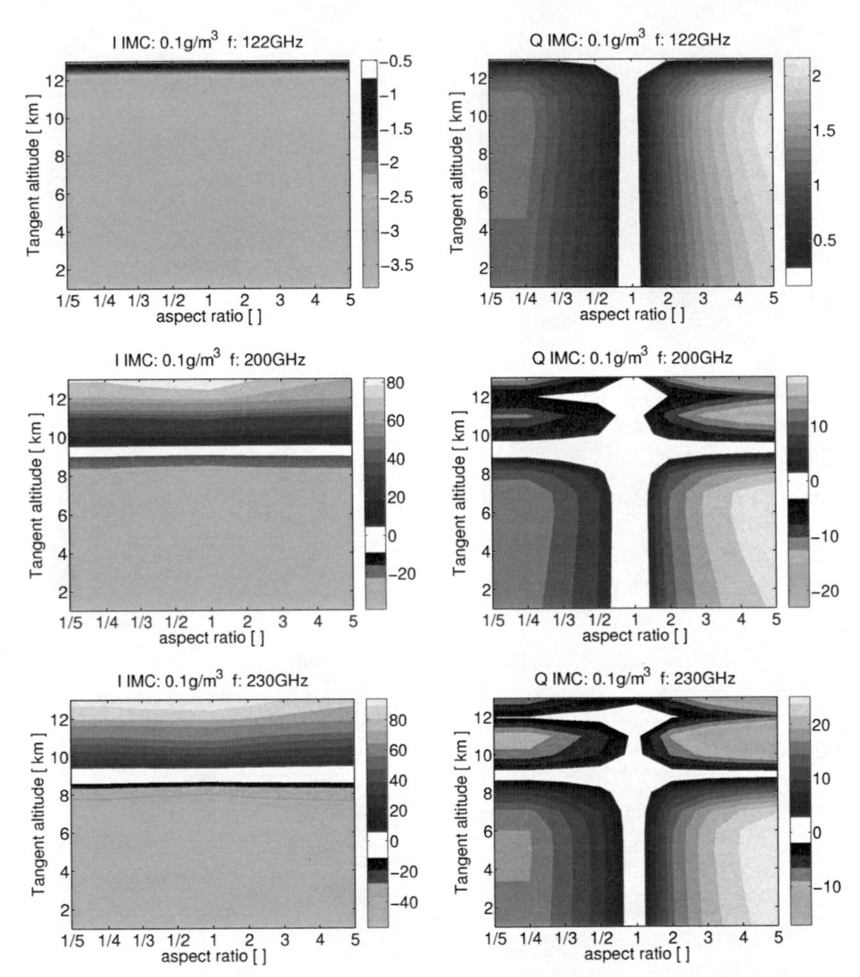

Figure 8.5: Dependence of total intensity and polarization on aspect ratio at 122 GHz, 200.5 GHz, and 230 GHz for a cloud with a constant IMC of 0.1 g/m^3. The contours correspond to the intensity difference compared to clear sky radiances (left) and to the polarization difference (right).

8.5 Combination of horizontally and vertically polarized channels

Since channel R1 (122 GHz) of the EOS MLS instrument measures both polarizations at the same time, it might be possible to use this channel to retrieve information about particle shape. Figure 8.6 shows scatter plots of the simulated radiances. The vertically polarized part of the radiation is plotted against the horizontally polarized part. The results for the particles with aspect ratio 1 (black circles) are on the diagonal, as they lead only to very small polarization. Different points of the same particle type correspond to different IMC. As we have also seen in Figure 8.5 the polarization difference is positive, which means that I_v is greater than I_h. The simulations show, that from the measurements it should be possible to gain information about particle shape. The further away the measurements are from the diagonal the higher is the deformation of the particles inside the cloud. However, it might be difficult to distinguish between oblate and prolate particles since both induce the same polarization state. The plot shows that for clouds with a rather large IMC the polarization difference in channel 122 GHz is sufficiently large to gain information about particle shape. The scatter plot looks similar for 4 km and for 11 km tangent altitude.

Since the scattering signal is much larger in channels R2 (200.5 GHz) and R3 (230 GHz), it should also be possible to obtain information about particle shape from those channels, since they measure different polarizations. Figure 8.7 shows scattered plots of R2 (vertical polarization) and R3 (horizontal polarization) at different tangent altitudes. Since the cloudy radiance is for tangent altitudes up to 9 km larger for R2 than for R3 (compare Figure 8.1), also the points for particles with aspect ratio 1 lie below the diagonal. The plots look similar to the plot for 122 GHz. With higher deformation the difference between R2 and R3 increases. Compared to R1A/B the scattering signal is much stronger, therefore the combination of R2 and R3 is more useful for measuring thin cirrus clouds. For tangent altitudes inside the cloud, it depends on the IMC, whether R2 or R3 measures larger BT. Comparing with Figure 8.5 it can be seen that the points above the diagonal result from the simulations for small IMC whereas the points below

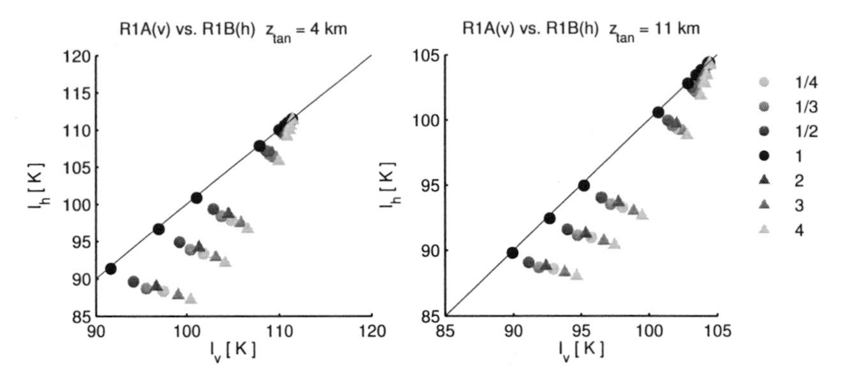

Figure 8.6: Scatter plot of vertically and horizontally polarized parts of the intensity for 122 GHz at tangent altitudes 4 and 11 km. This result corresponds to measurements of channel R1A (I_v) and R1B (I_h) of the EOS MLS instrument. Different symbols correspond to cloud particles of different aspect ratios.

the diagonal result from simulations for large IMC. Again simulations for more extreme aspect ratios are further away from the diagonal compared to simulations for less deformed particles. These results indicate that it might be possible to retrieve particle shape along with IMC from the measurements at tangent altitudes inside the cloud.

8.6 Conclusions and outlook

The first simulations for EOS MLS channels show, that the data, which will be obtained from the instrument, will be very useful to study cloud microphysics, like ice mass content and particle shape. Especially the different polarization characteristics can be used for this purpose. R1A/B (122 GHz) measure both polarizations for the same frequency at the same time, but the scattering signal is rather small at this frequency, so that this channel can probably only be used for studying rather thick clouds. The scattering signal in channels R2 (200.5 GHz) and R3 (230 GHz) is much larger, so that these channels can also be used for thin clouds. Unfortunately, R2 measures

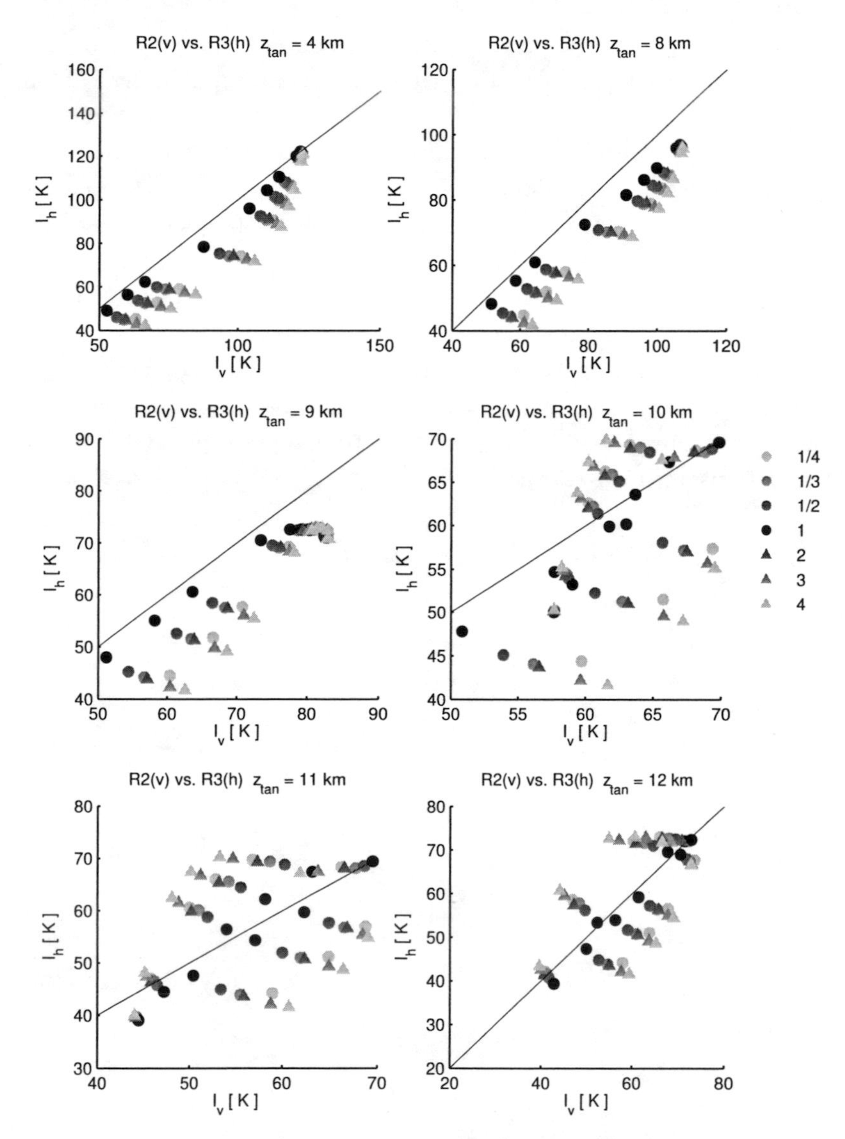

Figure 8.7: Scatter plots of vertically (200.5 GHz) and horizontally (230 GHz) polarized parts of the intensity at different tangent altitudes. This result corresponds to measurements of channel R2 (200.5 GHz, I_v) and R3 (230 GHz, I_h) of the EOS MLS instrument. Different symbols correspond to cloud particles of different aspect ratios.

only vertical polarization and R3 only horizontal polarization so that it is not possible to calculate directly the polarization difference Q for each channel. Nevertheless it is possible to combine the two channels in order to obtain cloud information, since absorption features and the scattering signal in the two channels are very similar. The possibility in retrieving cloud information in the higher frequency channels needs to be investigated. The problem is these channels might be, that the atmosphere is opaque, so that the instrument can not measure the clouds. The first data of the EOS MLS instrument is now being analyzed in the MLS science team at the Jet Propulsion Laboratory (JPL). The data will be available for other science teams in the near future. Then it will be possible to compare the DOIT model simulations with the real data.

9 Overall summary, conclusions and outlook

A new scattering algorithm, called Discrete Order ITerative (DOIT) method, was developed and implemented in the Atmospheric Radiative Transfer Simulator (ARTS). Before starting the development, literature related to radiative transfer modeling including a thermal source and scattering was reviewed. It turned out that none of the already existing models was well suited for the purpose of simulating polarized limb radiances in the microwave wavelength region, since all of the reviewed models use the plane-parallel approximation of the atmosphere. Some models using a spherical atmosphere were developed during the same time period as the DOIT algorithm, but these mostly use the scalar radiative transfer equation and other approximations. Besides the DOIT algorithm there is a Monte Carlo algorithm, which was also implemented into ARTS by Cory Davis in parallel to the implementation of the DOIT algorithm. The two algorithms are at present the only ones, which can model polarized limb radiances.

The basic equation of the scattering model is the vector radiative transfer equation (VRTE). The derivation of this equation requires basic principles and definitions of electromagnetic theory. There are several possibilities to calculate the scattering properties of small particles. The most simple method is the Rayleigh approximation for particles, which are very small compared to the wavelength of the radiation. For spherical particles and wavelengths comparable to the particle size, the Lorentz-Mie theory is usually applied. Since ice particles are of various shapes, which are mostly asymmetric, a more sophisticated method is required for modeling scattering of radiation in cirrus clouds. The T-matrix method, which is applicable for rotationally symmetric particles was chosen. Although the ice crystals are usually not rotationally symmetric, cylinders and plates are a good approximation for many of the crystals. There are methods for arbitrary particle shapes, but those methods are either computationally very expensive or not well tested. The T-matrix method is currently the most commonly used method for the calculation of scattering properties of cirrus clouds particles.

The VRTE is a matrix integro-differential equation which in general cannot be solved analytically. Several numerical methods have been invented to solve such kind of equations; these include discrete ordinate methods, Monte Carlo methods, and "doubling and adding" methods. The originality of the DOIT method is, that it is the first discrete ordinate algorithm for a spherical geometry. As a platform the ARTS clear sky model was used, as it includes modules for the calculation of gas absorption, for handling ray-tracing in a three-dimensional atmosphere and for the simulation of sensor characteristics. The DOIT algorithm solves the VRTE on a restricted part of the atmosphere denoted as the "cloud box", in order to minimize the computational effort. Briefly the algorithm can be described as follows: Scattering integrals, which are the difficult part of the VRTE, are first calculated at all cloud box grid points using the clear sky field. After that the VRTE can be solved using a fixed term for the scattering integral. The solutions for all cloud box points are the first iteration radiation field. Scattering integral fields and radiation fields are calculated alternately until convergence is obtained. In this way the VRTE is solved numerically for the cloud box. The spherical geometry of the cloud box required numerical optimizations, for instance the zenith angle grid optimization for the representation of the radiation field.

The 1D DOIT algorithm was compared to the model FM2D developed at RAL (Rutherford Appelton Laboratory) and the single scattering model KOPRA developed for MIPAS (Michelson Interferometer for Passive Atmospheric Sounding). ARTS-DOIT and the FM2D model showed excellent agreement (less than 1K difference in simulated brightness temperatures for most cloud cases) and ARTS-DOIT and KOPRA agreed well in the single scattering regime. KOPRA as well as FM2D neglect polarization. KOPRA only works for 1D spherical atmospheres, whereas the RAL model works in 1D and 2D pseudo-spherical atmospheres. The two models run faster than the ARTS model, but ARTS is the more general and more accurate model. The 3D polarized DOIT algorithm was compared to the ARTS Monte Carlo algorithm. This comparison was very important, since the two algorithms models are the very first ones, which are able to simulate polarized limb radiances in 3D spherical atmospheres for the microwave region. The agreement between the models was satisfactory. It shows that both the Monte Carlo method and the discrete ordinate method can be applied for solving the VRTE in a 3D spherical atmosphere.

Several simulation studies were performed using the new algorithm. The 1D scalar version was used to simulate frequency spectra for the MASTER instrument. The 1D polarized version was used for a sensitivity study of thin cirrus clouds on the EOS MLS instrument. Moreover, the 3D version was used for simulations of clouds with small horizontal extent. The results have shown that the effect of particle size is very significant on both intensity and polarization of the radiation. Particle shape is an important cloud parameter when the cloud particles are horizontally aligned with random azimuthal orientation. In the case of completely randomly oriented particles, changing the particle shape shows almost no effect in the simulations. For horizontally aligned particles, there is a significant difference between the scalar (unpolarized) version and the vector (polarized) version of the model in intensity. Therefore it is important to use a vector radiative transfer model to obtain accurate results, even if one is only interested in intensity, not in polarization. The 3D simulations show that one must not neglect cloud inhomogeneity effects. The scattering signal depends very much upon the sensor position with respect to the cloud. The fact that the scattering signal is much larger in limb geometry compared to down-looking geometries, due to the greater path-length through the cloud layers, demonstrates the potential of retrieving cloud properties from limb measurements.

ARTS is a modular program and can be run in different modes. The computation (CPU) time depends very much upon the chosen set-up, whether one uses the 1D- or the 3D-mode, or selects the polarized or the unpolarized mode. CPU time can also be reduced by calculating two instead of all four Stokes components. The accuracy of the results is not affected, as long as U and V are negligible. Grid optimization is very important for both accuracy and computation time. Overall, the 1D model, with or without polarization is rather efficient and can be used for example to calculate full frequency spectra. The 3D model however is very inefficient for larger 3D cloud domains, so that for such cases the Monte Carlo algorithm should be prefered. Although the 3D calculations are computationally demanding and therefore not yet useful for operational applications, the model is practical for research, for instance to study in detail the effect of different cloud parameters on polarization. A feature of the DOIT method is, that it yields the whole radiation field. To simulate radiances for different sensor positions, the radiation field only needs to be calculated once for the whole cloud box and the outgoing radiances can then be interpolated on each required viewing direction.

The ARTS package, which includes besides the scattering tools (Monte Carlo and DOIT) various functions for clear sky radiative transfer and sensor modeling, is freely available under the Gnu General Public License and can be downloaded from http://www.sat.uni-bremen.de/arts/.

In the near future data from the EOS MLS instrument will be available. This could be used first of all to validate the DOIT and Monte Carlo algorithms by testing whether they can simulate the real data. Later, after the development of a cloud retrieval algorithm, the DOIT model can be used as a forward model for cloud parameter retrievals from satellite data. The model will also be a crucial tool for the data analysis of the submillimeter limb sounder SMILES, which is planned to be launched in 2008.

Appendix

A Literature review

Before starting to develop a new scattering radiative transfer model a detailed literature review about already existing models was performed (Emde and Sreerekha, 2004). The models that were discussed in this review are summarized in Table A.1. Atmospheric geometries being used by the models and the applied methods for solving the radiative transfer equation are listed. Furthermore the table shows, which models can calculate the full Stokes vector, i.e., the polarization state of scattered radiances. The last column of the table inludes the particle types the models are able to handle.

Most of the reviewed models use a plane-parallel atmosphere. Only SHDOM, VDISORT and the Monte Carlo model, which are three-dimensional models, use an atmosphere which is discretized using a cartesian coordinate system. In these models the atmosphere does not consist just of plane-parallel layers, but of cuboidal grid cells. None of the models has used a spherical geometry which is necessary to simulate limb radiances. This is the major deficit. The new version of ARTS includes a spherical atmosphere.

Methods for 1D plane-parallel atmospheres are the Eddington approximations and the doubling-and-adding method. The advantage of the Eddington approximations is, that they give analytical expressions as solution and therefore the Eddington models are very fast, but they can only be applied for spherical particles. The doubling-and-adding method is a simple numerical method which is also quite fast and can be used for modeling all kinds of particle types.

The successive order of scattering method can be used for 1D and 3D atmospheres, in the 3D case in combination with the discrete ordinate method. For 3D calculations, Monte Carlo approaches are also possible. If one wants to calculate many viewing angles and different sensor positions for a rather small scattering domain, the discrete ordinate method is more efficient than the Monte Carlo method because the whole radiation field is calculated at once. On the other hand, if only a few viewing angles are needed, the Monte Carlo method is more efficient.

Microwave RT models for cloudy atmospheres

Name	Atmosphere	Method	Pol.	non-sph.
MWMOD Simmer (1993) Czekala (1999b)	plane-parallel	DO Iterative	y	y
RTTOV Eyre (1991) English and Hewison (1998)	plane-parallel	D-A, Eddington	n	n
SHDOM Evans (1998)	3D-cartesian	DO Iterative	n	n
PolRadTrans Evans and Stephens (1991)	plane-parallel	D-A	y	y
Hybrid Model Deeter and Evans (1998)	plane-parallel	Eddington & Single Scattering	n	n
DISORT Stamnes et al. (1988)	plane-parallel	DO Iterative	n	n
VDISORT Weng (1992) Schulz et al. (1999)	plane-parallel	DO Iterative	y	y
Perturbation Model Gasiewski and Stalin (1990)	plane-parallel	Perturbation method	y	y
Eddington Models Kummerow (1993)	plane-parallel	Eddington	n	n

Monte Carlo Roberti and Kummerow (1999), Liu et al. (1996)	3D-cartesian	Monte Carlo	y	y
VDOM Haferman et al. (1997)	3D-cartesian	DO Iterative	y	y
ARTS-MC Davis et al. (2004)	3D-spherical	Monte Carlo	y	y
ARTS-DOIT Emde et al. (2004a)	3D-spherical	DO Iterative	y	y

Table A.1: Overview of the reviewed radiative transfer models
Abbreviations: Pol.– polarization, DO – Discrete-ordinate method, D-A – Doubling-and-adding method, Single Scattering – Single scattering approximation, non-sph. – non-spherical particles

B Derivations

B.1 Solution of approximated VRTE

Equation (4.7) can be solved analytically using the following matrix exponential approach

$$\boldsymbol{I}^{(1)} = e^{-\overline{\langle \boldsymbol{K} \rangle} s} \boldsymbol{C}_1 + \boldsymbol{C}_2, \tag{B.1}$$

where $\boldsymbol{C}_1$ and $\boldsymbol{C}_2$ are constants which have to be determined. Substituting (B.1) into (4.7) gives the constant $\boldsymbol{C}_2$:

$$\begin{aligned} -\overline{\langle \boldsymbol{K} \rangle} e^{-\overline{\langle \boldsymbol{K} \rangle} s} \boldsymbol{C}_1 = & -\overline{\langle \boldsymbol{K} \rangle} e^{-\overline{\langle \boldsymbol{K} \rangle} s} \boldsymbol{C}_1 - \overline{\langle \boldsymbol{K} \rangle} \boldsymbol{C}_2 \\ & + \overline{\langle \boldsymbol{a} \rangle} \bar{B} + \overline{\left\langle \boldsymbol{S}^{(0)} \right\rangle} \\ \boldsymbol{C}_2 = & \overline{\langle \boldsymbol{K} \rangle}^{-1} \left(\overline{\langle \boldsymbol{a} \rangle} \bar{B} + \overline{\left\langle \boldsymbol{S}^{(0)} \right\rangle} \right). \end{aligned} \tag{B.2}$$

$\boldsymbol{C}_1$ can be determined using the initial condition, which is the radiation at the intersection point $\boldsymbol{P}'$ traveling towards the observation point $\boldsymbol{P}$:

$$\boldsymbol{I}^{(1)}(s = 0) = \boldsymbol{I}^{(0)}(\text{at intersection point}) \tag{B.3}$$

From the ansatz Equation (B.1) follows:

$$\begin{aligned} \boldsymbol{I}^{(0)} = & \boldsymbol{C}_1 + \overline{\langle \boldsymbol{K} \rangle}^{-1} \left(\overline{\langle \boldsymbol{a} \rangle} \bar{B} + \overline{\left\langle \boldsymbol{S}^{(0)} \right\rangle} \right) \\ \boldsymbol{C}_1 = & \overline{\boldsymbol{I}^{(0)}} - \overline{\langle \boldsymbol{K} \rangle}^{-1} \left(\overline{\langle \boldsymbol{a} \rangle} \bar{B} + \overline{\left\langle \boldsymbol{S}^{(0)} \right\rangle} \right) \end{aligned} \tag{B.4}$$

Substituting (B.2) and (B.4) into Equation (B.1) leads to the solution:

$$\begin{aligned} \boldsymbol{I}^{(1)} = & e^{-\overline{\langle \boldsymbol{K} \rangle} s} \cdot \left(\boldsymbol{I}^{(0)} - \overline{\langle \boldsymbol{K} \rangle}^{-1} \left(\overline{\langle \boldsymbol{a} \rangle} \bar{B} + \overline{\left\langle \boldsymbol{S}^{(0)} \right\rangle} \right) \right) \\ & + \overline{\langle \boldsymbol{K} \rangle}^{-1} \left(\overline{\langle \boldsymbol{a} \rangle} \bar{B} + \overline{\left\langle \boldsymbol{S}^{(0)} \right\rangle} \right) \end{aligned} \tag{B.5}$$

This can be resorted to the following form:

$$\boldsymbol{I}^{(1)} = e^{-\overline{\langle \boldsymbol{K} \rangle} s} \boldsymbol{I}^{(0)} + \left(\mathbb{I} - e^{-\overline{\langle \boldsymbol{K} \rangle} s} \right) \overline{\langle \boldsymbol{K} \rangle}^{-1} \left(\overline{\langle \boldsymbol{a} \rangle} \bar{B} + \overline{\left\langle \boldsymbol{S}^{(0)} \right\rangle} \right) \tag{B.6}$$

Here $\mathbb{I}$ denotes the identity matrix and $\boldsymbol{I}^{(0)}$ the Stokes vector at the intersection point. There are several ways to calculate the matrix exponential functions. In ARTS the Pade-approximation is implemented according to Moler and Loan (1979).

B.2 Transformation of single scattering properties from the particle frame to the laboratory frame for randomly oriented particles

B.2.1 Transformation from scattering matrix to phase matrix

Instead of calculating the phase matrix $\boldsymbol{Z}$, which relates the Stokes vectors relative to their respective meroidal planes, we can calculate the scattering matrix $\boldsymbol{F}$, which relates the Stokes parameters of the incident and the scattered beams with respect to the scattering plane. The scattering matrix for macroscopically isotropic and mirror-symmetric scattering media has only six independent matrix elements in contrast to the phase matrix which has in general sixteen independent matrix elements.

From symmetry considerations follows, that the scattering matrix (Equation (3.8)) has a simple block-diagonal structure. The advantages of using the particle frame are obvious: On the one hand side the calculation of the scattering matrix using the T-matrix method is very efficient and on the other hand side much less memory is required to store the phase matrix. It depends only on one angle instead of four and it has less elements. The only draw-back is, that a transformation from the particle frame to the laboratory frame is needed, as the radiative transfer calculations are performed in the laboratory frame.

Inserting the transformed $\boldsymbol{n}^{\mathrm{inc}}$ and $\boldsymbol{n}^{\mathrm{sca}}$ into Equation (3.1) we can calculate the scattering angle

$$\Theta = \arccos(\cos\theta^{\mathrm{sca}}\cos\theta^{\mathrm{inc}} + \sin\theta^{\mathrm{sca}}\sin\theta^{\mathrm{inc}}\cos(\phi^{\mathrm{sca}} - \phi^{\mathrm{inc}})) \quad \text{(B.7)}$$

where the angles θ^{sca} and ϕ^{sca} describe the scattered beam and θ^{inc} and ϕ^{inc} the incident beam in the laboratory frame.

For the transformation from the scattering matrix to the phase matrix different cases have to be considered:

1. For forward scattering ($\Theta = 0$) the scattering frame coincides with the laboratory frame and no transformation is required.

$$\boldsymbol{F} = \boldsymbol{Z} \quad \text{(B.8)}$$

2. Different transformations are needed depending of the difference between azimuth angles. We can derive the following transformation formulas for $0 < \phi^{\mathrm{sca}} - \phi^{\mathrm{inc}} < \pi$ provided that $\theta^{inc,sca}$ and $\phi^{inc,sca}$ are not equal to 0 or π:

$$\boldsymbol{Z}(\theta^{\mathrm{sca}}, \theta^{\mathrm{inc}}, \phi^{\mathrm{sca}}, \phi^{\mathrm{inc}}) =$$

$$\begin{bmatrix} F_{11} & C_1F_{12} & S_1F_{12} & 0 \\ C_2F_{12} & C_1C_2F_{22} - S_1S_2F_{33} & S_1C_2F_{22} + C_1S_2F_{33} & S_2F_{34} \\ -S_1F_{12} & -C_1S_2F_{22} - S_1C_2F_{33} & -S_1S_2F_{22} + C_1C_2F_{33} & C_2F_{34} \\ 0 & S_2F_{34} & -C_1F_{34} & F_{44} \end{bmatrix} \quad \text{(B.9)}$$

where

$$C_j = \cos 2\sigma_j = 2\cos^2\sigma_j - 1 \quad \text{(B.10)}$$

$$S_j = \sin 2\sigma_j = 2\sqrt{1 - \cos^2\sigma_j}\cos\sigma_j \quad \text{(B.11)}$$

$$j = 1, 2$$

The terms $\cos\sigma_1$ and $\cos\sigma_2$ can be calculated from θ^{sca}, ϕ^{sca}, θ^{inc} and ϕ^{inc} using spherical trigonometry:

$$\cos\sigma_1 = \frac{\cos\theta^{\mathrm{sca}} - \cos\theta^{\mathrm{inc}}\cos\theta}{\sin\theta^{\mathrm{inc}}\sin\theta} \quad \text{(B.12)}$$

$$\cos\sigma_2 = \frac{\cos\theta^{\mathrm{inc}} - \cos\theta^{\mathrm{sca}}\cos\theta}{\sin\theta^{\mathrm{sca}}\sin\theta} \quad \text{(B.13)}$$

For different azimuth angles, the formulas look very similar. Only some signs are changed.

3. In the case that $\theta^{inc,sca}$ or $\phi^{inc,sca}$ equal zero or π, the above formulas are not defined. The limiting values can be derived:

$$\begin{aligned}
\lim_{\theta^{\text{sca}}\to 0} \cos\sigma_2 &= -\cos(\phi^{\text{sca}} - \phi^{\text{inc}}) \\
\lim_{\theta^{\text{sca}}\to \pi} \cos\sigma_2 &= \cos(\phi^{\text{sca}} - \phi^{\text{inc}}) \\
\lim_{\theta^{\text{inc}}\to 0} \cos\sigma_1 &= -\cos(\phi^{\text{sca}} - \phi^{\text{inc}}) \\
\lim_{\theta^{\text{inc}}\to \pi} \cos\sigma_1 &= \cos(\phi^{\text{sca}} - \phi^{\text{inc}})
\end{aligned} \tag{B.14}$$

4. For backward scattering ($\Theta = \pi$) the scattering matrix is diagonal and has only two independent elements:

$$\boldsymbol{F}(\pi) = \begin{bmatrix} F_{11}(\pi) & 0 & 0 & 0 \\ 0 & F_{22}(\pi) & 0 & 0 \\ 0 & 0 & -F_{22}(\pi) & 0 \\ 0 & 0 & 0 & F_{11}(\pi) - F_{22}(\pi) \end{bmatrix} \tag{B.15}$$

As the phase matrix the scattering matrix of course also depends on the frequency and on the temperature or equivalently on the refractive index of the scattering medium.

B.2.2 Extinction matrix and absorption vector

For scattering media consisting of randomly oriented particles one can show, that all off-diagonal elements of the extinction matrix $\boldsymbol{K}$ vanish. Furthermore, all diagonal elements are equal and correspond to the extinction cross-section C_{ext}.

$$\boldsymbol{K} = C_{\text{ext}}\mathbb{I} = N \langle C_{\text{ext}} \rangle \mathbb{I} \tag{B.16}$$

where $\mathbb{I}$ is the identity matrix, N the number of particles in a volume element and $\langle C_{\text{ext}} \rangle$ the average extinction cross-section per particle, which in this case is independent of the direction of propagation and of the polarization state of the incident radiation.

The ensemble-averaged emission vector for isotropic scattering media must be independent of the emission direction. It can be shown, that the absorption vector just depends on the absorption cross-section C_{abs}

$$\boldsymbol{a} = \begin{bmatrix} C_{\mathrm{abs}} \\ 0 \\ 0 \\ 0 \end{bmatrix} = \begin{bmatrix} N \langle C_{\mathrm{abs}} \rangle \\ 0 \\ 0 \\ 0 \end{bmatrix} \tag{B.17}$$

where $\langle C_{\mathrm{abs}} \rangle$ is the average absorption cross-section per particle.

Absorption end emission cross-section depend on frequency and on the refractive index being a function of temperature.

Acknowledgments

First of all, I would like to express my sincere thanks to my supervisor Stefan Bühler for his continued interest, encouragement, support and valuable help during the accomplishment of this work, and for good team work in the ARTS development. He has implemented functions to calculate gas absorption, which are used in the DOIT algorithm. Furthermore I would like to thank my colleagues from the SAT group for the very nice working atmosphere and also for the useful scientific discussions. Special thanks goes to Sreerekha T. R. for the good collaboration during the development and testing phase of the DOIT model and for the useful scientific discussions. I also appreciate the work of Claas Teichmann, who has used and tested the new model for his diploma thesis work on polarization. In addition I would like to thank Oliver Lemke for programming assistance and technical support.

Next, I would like to thank Cory Davis from the University of Edinburgh, who has developed the Monte Carlo scattering model. Since many parts of the models are in common, our collaboration was very expedient. In addition, I would like to thank Patrick Eriksson from the Chalmers Institute of Technology in Gothenburg, who is also part of the ARTS development team, for suggestions and ideas. He has implemented the 3D ray-tracing scheme, which is used in the DOIT algorithm.

A large fraction of the work was carried out in the context of an ESTEC study (Contract No. 15457/02/NL/MW). During this study, the comparisons of the DOIT model with the models FM2D and KOPRA were carried out. I appreciate very much the collaboration with Richard Siddans from the the Rutherford Appleton Laboratory (RAL) near Oxford, who has implemented scattering in the model FM2D and who helped me finding several bugs in the initial DOIT algorithm. Furthermore, I would like to thank Michael Höpfner from the "Forschungszentrum" in Karlsruhe, who did a lot of work for the comparison study between the DOIT model and the KOPRA model, in which he has included a scattering algorithm. I also thank all other members of the consortium for the good team work. I very much appreciate suggestions and comments from Brian Kerridge (RAL), who was the supervisor of the science consortium, and from Jörg Langen from ESTEC, whose criticism was always well founded and helpful.

For the calculation of single scattering properties, many different public domain programs were used. I would like to thank Michael Mishchenko and Steven Warren for making available the T-matrix program and the refractive index program, respectively. Furthermore I thank Christian Mätzler for providing the Mie program.

I would like to express my gratitude to Prof. Klaus Künzi and Prof. Clemens Simmer for reviewing this thesis and for helpful comments and suggestions.

Last but not least I would like to thank my parents and my friends, who were always there when I needed them. Special thanks goes to Bruno Matzas for reading and commenting on the manuscript.

Besides the ESTEC study mentioned earlier, this work was funded by the German Federal Ministry of Education and Research (BMBF), within the DLR project SMILES, grant 50 EE 9815, and within the AFO2000 project UTH-MOS, grant 07ATC04. It is also a contribution to the COST Action 723 'Data Exploitation and Modeling for the Upper Troposphere and Lower Stratosphere'.

C List of acronyms

Acronym	Meaning
ARTS	Atmospheric Radiative Transfer Simulator
BT	Brightness Temperature
CLEAS	Cryogenic Limb Array Etalon Spectrometer
CPU	Central Processing Unit
CRISTA	Cryogenic Spectrometers and Telescopes for the Atmosphere
DDA	Discrete Dipole Approximation
DDSCAT	Discrete Dipole Approximation for Scattering and Absorption of Light by Irregular Particles
DISORT	Discrete Ordinate Radiative Transfer Model
DOIT	Discrete Ordinate ITerative method
DOM	Discrete Ordinate Method
ECBM	Extended Boundary Condition Method
EOS	Earth Observing System
ESA	European Space Agency
ESTEC	European Space research and TEchnology Centre
FASCOD	Fast Atmosphere Signature Code
FIRE	First ISCCP Regional Experiment
FM2D	Forward Model 2D
GCM	Global Climate Models
GOME	Global Ozone Monitoring Experiment
GOMETRAN	GOME radiative TRANsfer model
HITRAN	High-resolution Transmission Molecular Absorption database

Acronym	Meaning
ISCCP	International Satellite Cloud Climatology Project
IMC	Ice Mass Content
JPL	Jet Propulsion Laboratory
KOPRA	Karlsruhe Optimized and Precise Radiative Transfer Algorithm
LOS	Line Of Sight
MATLAB	MAtrix LABoratory
MAS	Millimeter Atmospheric Sounder
MASTER	Millimeter Wave Acquisitions for Strato-sphere/Troposphere Exchange Research
MC	Monte Carlo method
MIPAS	Michelson Interferometer for Passive Atmo-spheric Sounding
MLS	Microwave Limb Sounder
MWMOD	MicroWave MODel
NASA	National Aeronautics and Space Admisistration
PyARTS	Python ARTS
RAL	Rutherford Appleton Laboratory
RT	Radiative Transfer
RTTOV	fast Radiative Transfer model for TOVs
SHDOM	Spherical Harmonics Discrete Ordinate Method
SMILES	Superconduction Submillimeter-Wave Limb Emission Sounder
SMR	Submillimeter Radiometer
SRTE	Scalar Radiative Transfer Equation
SWCIR	Submillimeter-Wave Cloud Ice Radiometer
TES	Troposhperic Emission Spectrometer
TOVS	Tiros Operational Vertical Sounder
UARS	Upper Atmospheric Research Satellite
VDISORT	Vector Discrete Ordinate Radiative Transfer Model
VDOM	Vector Discrete-Ordinates Method
VRTE	Vector Radiative Transfer Equation

D List of symbols

Symbol	Definition and dimension in SI units	Introduced in section
$\boldsymbol{a}^p$	particle absorption vector [m^2]	1.3.4
$\langle \boldsymbol{a} \rangle$	total ensemble averaged absorption vector (includes particle and gas contributions) [m^{-1}]	1.5
$\langle \boldsymbol{a}^g \rangle$	averaged gas absorption vector [m^{-1}]	1.5
$\langle \boldsymbol{a}^p \rangle$	ensemble averaged particle absorption vector [m^{-1}]	1.5
$\langle \boldsymbol{a}_i^p \rangle$	ensemble averaged absorption vector for one particle type [m^2]	1.5
B	Planck blackbody energy distribution [W s m^{-2} sr^{-1}]	1.3.4
c	speed of light in vacuum [m s^{-1}]	1.1
C_{abs}	absorption cross section [m^2]	1.3.5
C_{ext}	extinction cross section [m^2]	1.3.5
C_{sca}	scattering cross section [m^2]	1.3.5
$\boldsymbol{E}$	electric field vector [V m^{-1}]	1.1
$\boldsymbol{E}_0$	amplitude of electric field vector [V m^{-1}]	1.1
E_θ, E_ϕ	spherical coordinate components of the electric field vector [V m^{-1}]	1.2
E_h, E_v	horizontal and vertical components of the electric field vector [V m^{-1}]	1.2
$\boldsymbol{F}$	scattering matrix [m^2]	3.3
g	probability density function [–]	5.3
$\boldsymbol{H}$	magnetic field vector [A m^{-1}]	1.1

Symbol	Definition and dimension in SI units	Introduced in section
$\hbar$	Planck constant divided by 2π [Js]	1.3.4
I	intensity, first Stokes parameter [W m^{-1}]	1.2
I	specific intensity [W s m^{-2} sr^{-1}]	1.5
$\boldsymbol{I}$	Stokes vector [W m^{-1}]	1.2
$\boldsymbol{I}$	specific intensity vector or "Stokes vector" [W s m^{-2} sr^{-1}]	1.5
$\boldsymbol{I}_b$	blackbody Stokes column vector [W s m^{-2} sr^{-1}]	1.3.4
$\mathcal{I}$	radiation field [W s m^{-2} sr^{-1}]	4.1.1
$\mathcal{I}^{(n)}$	nth order radiation field [W s m^{-2} sr^{-1}]	4.1.1
IMC	ice mass content [kg m^{-3}]	3.5
$k = k_{\mathrm{R}} + k_{\mathrm{I}}$	(complex) wave number [m^{-1}]	1.1
k_b	Boltzmann constant [JK^{-1}]	1.3.4
$\boldsymbol{K}$	total extinction matrix [m^{-1}]	1.3.3
$\langle \boldsymbol{K} \rangle$	ensemble averaged total extinction matrix [m^{-1}]	1.5
$\langle \boldsymbol{K}^g \rangle$	ensemble averaged gaseous extinction matrix [m^{-1}]	1.5
$\langle \boldsymbol{K}^p \rangle$	ensemble averaged particle extinction matrix [m^{-1}]	1.5
$\langle \boldsymbol{K}_i^p \rangle$	ensemble averaged extinction matrix for one particle type [m^2]	1.5
$m = m_{\mathrm{R}} + m_{\mathrm{I}}$	(complex) refractive index relative to vacuum of surrounding medium [–]	1.1
m	mass of a particle [kg]	3.5
N	number of particles [–]	1.4
n^g	volume mixing ratio [–]	1.5
n^p	particle number density [m^{-3}]	1.5
$n(r)$	particle size distribution function [–]	3.5
$\hat{\boldsymbol{n}}$	unit vector [–]	1.2
$\hat{\boldsymbol{n}}^{\mathrm{inc}}$	unit vector in the incidence direction [–]	1.3

Symbol	Definition and dimension in SI units	Introduced in section
$\hat{\boldsymbol{n}}^{\mathrm{sca}}$	unit vector in the scattering direction [–]	1.3
p	phase function [–]	1.3.5
p	degree of polarization [–]	1.2
p_{lin}	degree of linear polarization [–]	1.2
p_{circ}	degree of circular polarization [–]	1.2
$\mathbf{P}$	time-averaged Poynting vector [W m^{-2}]	1.1
$\vec{p}$	pressure grid [Pa]	4.1.1
Q	second Stokes parameter [W m^{-1}]	1.2
Q	second component of specific intensity vector [W s m^{-2} sr^{-1}]	1.5
Q_{abs}	absorption efficiency [–]	3.3
Q_{ext}	extinction efficiency [–]	3.3
Q_{sca}	scattering efficiency [–]	3.3
r	distance from the origin of a coordinate system [m]	1.3
r	equal volume sphere radius of a particle [m]	3.2
$\tilde{r}$	random number [–]	5.3
$\boldsymbol{r}$	radius (position) vector [m]	1.1
R_{eff}	effective radius of a particle size distribution [m]	3.5
R_{me}	median radius of a particle size distribution	5.1
$\langle S \rangle$	scattering source function [W s m^{-1} sr^{-1}]	5.1
$\mathcal{S}^{(n)}$	nth order scattering integral field [W s m^{-2} sr^{-1}]	4.1.1
t	time [s]	1.1
T	temperature [K]	1.3.4
T_{Planck}	Planck brightness temperature [K]	2.7
T_{RJ}	Rayleigh Jeans brightness temperature [K]	2.7

Symbol	Definition and dimension in SI units	Introduced in section
U	third Stokes parameter [W m^{-1}]	1.2
U	third component of specific intensity vector [W s m^{-2} sr^{-1}]	1.5
V	fourth Stokes parameter [W m^{-1}]	1.2
V	fourth component of specific intensity vector [W s m^{-2} sr^{-1}]	1.5
V	volume of a particle [m^{-3}]	3.5
W	power [W]	1.3.4
x	scattering parameter [–]	3.3
$\boldsymbol{Z}$	phase matrix [m^2]	1.3.2
$\langle \boldsymbol{Z} \rangle$	ensemble averaged phase matrix [m^{-1}]	1.4
$\langle \boldsymbol{Z}_i \rangle$	ensemble averaged phase matrix for one particle type [m^2]	1.5
α	polarizability [m^3]	3.3
α_i^g	individual gas absorption coefficient [m^{-1}]	1.5
α^p	particle absorption coefficient [m^{-1}]	1.1
$\langle \alpha^g \rangle$	averaged gas absorption coefficient [m^{-1}]	1.5
$\vec{\alpha}$	latitude grid [°]	4.1.1
$\vec{\beta}$	longitude grid [°]	4.1.1
Δs	path length element [m]	5.1
ΔS	surface element [m^2]	1.3.3
$\Delta\omega$	angular frequency interval [s^{-1}]	1.3.4
$\Delta\Omega$	solid angle [sr]	1.3.4
Δ_n	phase of amplitude matrix [–]	1.4
ϵ	electric permittivity [F m^{-1}]	1.1
λ	free space wavelength [m]	1.1
μ	magnetic permeability [H m^{-1}]	1.1
ν	frequency of radiation [s^{-1}]	1.1
ω	angular frequency [s^{-1}]	1.1
ω_0	single scattering albedo [–]	1.3.5

Symbol	Definition and dimension in SI units	Introduced in section
$\tilde{\omega}$	similar to scattering albedo, used in Monte Carlo model [–]	5.3
π	pi [–]	1.5
τ_{max}	maximal optical depth [–]	4.1.4
$\vec{\phi}$	azimuth angle grid [°]	4.1.1
ρ	density of a scattering medium [gm^{-3}]	3.5
$\vec{\theta}$	zenith angle grid [°]	4.1.1
Θ	scattering angle [°]	3.2

Symbol	Definition and dimension in SI units	Introduced in section
	General notation	
x^*	complex conjugate of x	1.1
$\langle x \rangle$	ensemble average of x	1.4
$\boldsymbol{X}$	matrix $\boldsymbol{X}$	1.1
X_{ij}	element (ij) of $\boldsymbol{X}$	1.3.2
$\boldsymbol{x}$	vector $\boldsymbol{x}$	1.1
x_i	i^{th} element of $\boldsymbol{x}$	1.5
x^{inc}	x for incident direction	1.3.1
x^{sca}	x for scattering direction	1.3.1
$\boldsymbol{X}^T$	transpose of $\mathbf{X}$	1.5
$\int_{4\pi}$	integral over the whole space	1.5
$\text{Re}\{\text{x}\}$	Real part of x	1.1
$\text{Im}\{\text{x}\}$	Imaginary part of x	1.1
$\overline{x}$	spatial average of x	2.3.3
Γ	gamma function	3.5
$\mathbb{I}$	identity matrix	B.1

Bibliography

Anderson, G. P., Clough, S. A., Kneizys, F. X., Chetwynd, J. H. and Shettle, E. P., 1986: *AFGL atmospheric constituent profiles (0–120 km)*. Tech. Rep. TR-86-0110, AFGL.

Arking, A., 1991: The radiative effects of clouds and their impact on climate. *Bull. Am. Meteorol. Soc.*, **72**, 795–813.

Beer, R., Glavich, T. A. and Rider, D. M., 2001: Tropospheric emission spectrometer for the Earth Observing System's Aura satellite. *Appl. Opt.*, **40**, 15, 2356–2367.

Bohren, C. and Huffman, D. R., 1998: *Absorption and Scattering of Light by Small Particles.* Wiley Science Paperback Series.

Buehler, S. A., 1999: *Microwave Limb Sounding of the Stratosphere and Upper Troposphere.* Berichte aus der Physik, PO-Box 1290, D 52013 Aachen: Shaker Verlag GmbH. PhD thesis, University of Bremen, ISBN 3-8265-4745-4, 262 pages.

Buehler, S. A., Eriksson, P., Kuhn, T., von Engeln, A. and Verdes, C., 2005a: ARTS, the Atmospheric Radiative Transfer Simulator. *J. Quant. Spectrosc. Radiat. Transfer*, **91**, 1, 65–93.

Buehler, S. A., Verdes, C. L., Tsujimaru, S., Kleinboehl, A., Bremer, H., Sinnhuber, M. and Eriksson, P., 2005b: The Expected Performance of the SMILES Submillimeter-Wave Limb Sounder compared to Aircraft Data. *Radio Sci.* In press.

von Clarmann, T., Dudhia, A., Edwards, D. P., Funke, B., Höpfner, M., Kerridge, B., Kostsov, V., Linden, A., López-Puertas, M. and Timofeyev, Y. M., 2002: Intercomparison of radiative transfer codes under non-local thermodynamic equilibrium conditions. *J. Geophys. Res.*, **107**, D22, 4631, doi:10.1029/2001JD001 551.

von Clarmann, T., Höpfner, M., Funke, B., López-Puertas, M., Dudhia, A., Jay, V., Schreier, F., Ridolfi, M., Ceccherini, S., Kerridge,

B. J., Reburn, J. and Siddans, R., 2003: Modelling of atmospheric mid–infrared radiative transfer: the AMIL2DA algorithm intercomparison experiment. *J. Quant. Spectrosc. Radiat. Transfer*, **78**, 3-4, 381–407 doi:10.1016/S0022–4073(02)00 262–5.

Czekala, H., 1999a: *Microwave radiative transfer calculations with multiple scattering by nonspherical hydrometeors.* Ph.D. thesis, Rheinische Friedrich-Wilhems-Universität Bonn, Auf dem Hügel 20, 53121 Bonn.

Czekala, H., 1999b: *Microwave radiative transfer calculations with multiple scattering by nonspherical hydrometeors.* Ph.D. thesis, Rheinische Friedrich-Wilhelms-Universitaet Bonn, Auf dem Huegel 20, 53121 Bonn, Germany.

Czekala, H. and Simmer, C., 1998: Microwave radiative transfer with nonspherical precipitating hydrometeors. *J. Quant. Spectrosc. Radiat. Transfer*, **60**, 3, 365–374.

Davis, C., Emde, C. and Harwood, R., 2004: A 3D Polarized Reversed Monte Carlo Radiative Transfer Model for mm and sub-mm Passive Remote Sensing in Cloudy Atmospheres. *IEEE T. Geosci. Remote.* In press.

Deeter, M. N. and Evans, K. F., 1998: A hybrid Eddington-single scattering radiative transfer model for computing radiances from thermally emitting atmospheres. *J. Quant. Spectrosc. Radiat. Transfer*, **60**, 4, 635–648.

Del Genio, A. D., Wolf, A. B., Mace, G. G. and Miloshevich, L. M., 2002: Observed Regimes of Mid-latitude and Tropical Cirrus Mircophysical Behavior. In *Twelfth ARM Science Team Meeting Proceedings*, St. Petersburg, Florida.

Donovan, D. P., 2003: Ice-cloud effective particle size parametrization based on combined lidar, radar reflectivity, and mean Doppler velocity measurements. *J. Geophys. Res.*, **108**, D18, 4573. Doi:10.1029/2002JD003469.

Draine, B. T. and Flateau, P. J., 2003: *User Guide to the Discrete Dipole Approximation Code DDSCAT.6.0.* URL http://arxiv.org/abs/astro-ph/0300969.

Emde, C., Buehler, S. A., Davis, C., Eriksson, P., Sreerekha, T. R. and Teichmann, C., 2004a: A Polarized Discrete Ordinate Scattering

Model for Simulations of Limb and Nadir Longwave Measurements in 1D/3D Spherical Atmospheres. *J. Geophys. Res.*, **109**, D24.

Emde, C., Buehler, S. A., Eriksson, P. and Sreerekha, T. R., 2004b: The effect of cirrus clouds on microwave limb radiances. *J. Atmos. Res.*, **72**, 1–4, 383–401.

Emde, C. and Sreerekha, T. R., 2004: *Development of a RT model for frequencies between 200 and 1000 GHz, WP1.2 Model Review.* Tech. Rep., ESTEC Contract No AO/1-4320/03/NL/FF.

English, S. and Hewison, T., 1998: A fast generic millimeter wave emissivity model. In *Microwave Remote Sensing of the Atmosphere and the Environment*, T. Hayasaka, Y. J., D.L. Wu and Jiang, J., eds., vol. 3503 of *Proceedings of SPIE*, SPIE.

Eriksson, P. and Buehler, S. A. (Eds.), 2001: *Atmospheric Millimeter and Sub-Millimeter Wave Radiative Transfer Modeling II.* Logos Verlag Berlin. ISBN 3-89722-585-9, ISBN 1615-6862.

Eriksson, P., Buehler, S. A., Emde, C., Sreerekha, T. R., Melsheimer, C. and Lemke, O., 2004: *ARTS-1-1 User Guide.* University of Bremen. 308 pages, regularly updated versions available at www.sat.uni-bremen.de/arts/.

Eriksson, P., Merino, F., Murtagh, D., Baron, P., Ricaud, P. and de la Noë, J., 2002: Studies for the Odin sub-millimetre radiometer: 1. Radiative transfer and instrument simulation. *Can. J. Phys.*, **80**, 321–340.

Evans, K. F., 1998: The spherical harmonics disrete ordinate method for three-dimensional atmospheric radiative transfer. *J. Atmos. Sci.*, **55**, 429–466.

Evans, K. F. and Stephens, G. L., 1991: A new polarized atmospheric radiative transfer model. *J. Quant. Spectrosc. Radiat. Transfer*, **46**, 5, 412–423.

Evans, K. F., Walter, S. J., Heymsfield, A. J. and Deeter, M. N., 1998: Modeling of Submillimeter Passive Remote Sensing of Cirrus Clouds. *J. Appl. Met.*, **37**, 184–205.

Evans, K. F., Walter, S. J., Heymsfield, A. J. and McFarquhar, G. M., 2002: Submillimeter-Wave Cloud Ice Radiometer: Simulations of retrieval algorithm performance. *J. Geophys. Res.*, **107**, D3, 10.1029/2001JD000 709.

Eyre, J. R., 1991: *A fast radiative transfer model for satellite sounding systems.* Tech. Rep., ECMWF Tech. Memo. No. 176.

Fischer, H. and Oelhaf, H., 1996: Remote sensing of vertical profiles of atmospheric trace constituents with MIPAS limb-emission spectrometers. *Appl. Opt.*, **35**, 16, 2787–2796.

Gasiewski, A. J. and Stalin, D. H., 1990: Numerical modeling of passive microwave O_2 observations over precipitation. *Radio Science*, **25**, 3, 217–235.

Glatthor, N., Höpfner, M., Stiller, G. P., von Clarmann, T., Dudhia, A., Echle, G., Funke, B. and Hase, F., 1999: Intercomparison of the KOPRA and the RFM Radiative Transfer Codes. In *Proc. European Symposium on Atmospheric Measurements from Space, ESAMS'99, 18–22 Jan 1999, Noordwijk*, European Space Agency, ESTEC, Noordwijk, The Netherlands.

Haferman, J. L., Smith, T. F. and Krajewski, W. F., 1997: A multidimensional discrete-ordinates method for polarised radiative transfer. Part 1: Validation for randomly oriented axisymmetric particles. *J. Quant. Spectrosc. Radiat. Transfer*, **58**, 379–398.

Hartmann, G. K., Bevilacqua, R. M., Schwartz, P. R., Kämpfer, N., Kuenzi, K. F., Aellig, C. P., Berg, A., Boogaerts, W., Connor, B. J., Croskey, C. L., Daehler, M., Degenhardt, W., Dicken, H. D., Goldizen, D., Kriebel, D., Langen, J., Loidl, A., , Olivero, J. J., Pauls, T. A., Puliafito, S. E., Richards, M. L., Rudin, C., Tsou, J. J., Waltman, W. B., Umlauft, G. and Zwick, R., 1996: Measurements of O_3 , H_2O, and ClO in the middle atmosphere using the Millimeter–wave Atmospheric Sounder (MAS). *Geophys. Res. Lett.*, **23**, 17, 2313–2316.

Havemann, S. and Baran, A. J., 2001: Extension of T-matrix to scattering of electromagnetic plane waves by non-axisymmetric dielectric particles: application to hexagonal ice cylinders. *J. Quant. Spectrosc. Radiat. Transfer*, **70**, 2, 139–158.

Heymsfield, A. J. and Platt, C. M. R., 1984: A parametrization of the particle size spectrum of ice clouds in terms of the ambient temperature and the ice water content. *J. Atmos. Sci.*, **41**, 864–855.

Hoepfner, M. and Emde, C., 2005: Comparison of single and multiple scattering approaches for the simulation of limb-emission observations in the mid-IR. *J. Quant. Spectrosc. Radiat. Transfer*, **91**, 3, 275–285.

Ishimoto, H. and Masuda, K., 2002: A Monte Carlo approach for the calculation of polarized light: application to an incident narrow beam. *J. Quant. Spectrosc. Radiat. Transfer*, **72**, 4, 462–483.

Jackson, J. D., 1998: *Classical electrodynamics.* New York: John Wiley & Sons.

Kerridge, B., Jay, V., Reburn, J., Siddans, R., Latter, B., Lama, F., Dudhia, A., Grainger, D., Burgess, A., Höpfner, M., Steck, T., Stiller, G., Buehler, S., Emde, C., Eriksson, P., Ekström, M., Baran, A. and Wickett, M., 2004: *Consideration of mission studying chemistry of the UTLS, Task 2 Report.* Tech. Rep., ESTEC Contract No 15457/01/NL/MM.

Kinne, S. et al., 1997: Cirrus cloud radiative and microphysical properties from ground observations and in situ measurements during FIRE 1991 and their applications to exhibit problems in cirrus solar radiative transfer modeling. *Journal of Atmospheric Science*, **54**, 2320–2344.

Kummerow, C., 1993: On the accuracy of the Eddington Approximation for radiative transfer in the microwave frequencies. *J. Geophys. Res.*, **98**, D2, 2757–2765.

Liou, K. N., 2002: *An Introduction to Atmospheric Radiation*, vol. 84 of *International Geophysics Series.* Academic Press, Elsevier Science.

Liou, K. N., Yang, P., Takano, Y., Sassen, K., Charlock, T. P. and Arnott, W. P., 1998: On the radiative properties of contrail cirrus. *Geophys. Res. Lett.*, **25**, 1161-1164.

Liu, Q., Simmer, C. and Ruprecht, E., 1996: Three-dimensional radiative transfer effects of clouds in the microwave spectral range. *J. Geophys. Res.*, **101**, D2, 4289–4298.

Mätzler, C., 1998: *Solar System Ices*, vol. 227 of *Astrophys. and Space Sci. Library*, chap. Microwave properties of ice and snow. Dordrecht: Kluwer Academic publishers, 241–257.

Mätzler, C., 2002: *MATLAB Functions for Mie Scattering and Absorption - Version 2.* Tech. Rep. 2002-11, Universität Bern.

McFarquhar, G. M. and Heymsfield, A. J., 1997: Parametritation of Tropical Cirrus Ice Crystal Size Distributions and Implications for Radiative Transfer: Results from CEPEX. *J. Atmos. Sci.*, **54**, 2187–2200.

Mishchenko, M. I., 2000: Calculation of the amplitude matrix for a nonspherical particle in a fixed orientation. *Applied Optics*, **39**, 1026–1031.

Mishchenko, M. I., 2002: Vector radiative transfer equation for arbitrarily shaped and arbitrarily oriented particles: a microphysical derivation from statistical electromagnetics. *Appl. Opt.*, **41**, 7114–7134.

Mishchenko, M. I., Hovenier, J. W. and Travis, L. D. (Eds.), 2000: *Light Scattering by Nonspherical Particles.* Academic Press, ISBN 0-12-498660-9.

Mishchenko, M. I., Travis, L. D. and Lacis, A. A., 2002: *Scattering, Absorption, and Emission of Light by Small Particles.* Cambridge University Press.

Moler, C. B. and Loan, C. F. V., 1979: Nineteen Dubious Ways to Compute the Exponential of a Matrix. *SIAM Review*, **20**, 801–836.

Murtagh, D., Frisk, U., Merino, F., Ridal, M., Jonsson, A., Stegman, J., Witt, G., Eriksson, P., Jimenez, C., Megie, G., de la Noe, J., Ricaud, P., Baron, P., Pardo, J. R., Hauchcorne, A., Llewellyn, E. J., Degenstein, D. A., Gattinger, R. L., Lloyd, N. D., Evans, W. F. J., McDade, I. C., Haley, C. S., Sioris, C., von Savigny, C., Solheim, B. H., McConnell, J. C., Strong, K., Richardson, E. H., Leppelmeier, G. W., Kyro, E., Auvinen, B. H. and Oikarinen, L., 2002: An overview of the Odin atmospheric mission. *Can. J. Phys.*, **80**, 309–319, URL www.nrc.ca/cgi-bin/cisti/journals/rp/rp2_abst_e?cjp_p01-157_8%0_ns_nf_cjpS1-02.

Oikarinen, L., Sihvola, E. and Kyrola, E., 1999: Multiple scattering radiance in limb-viewing geometry. *J. Geophys. Res.*, **104**, D24, 31 261–31 274.

Roberti, L., Haferman, J. and Kummerow, C., 1994: Microwave radiative transfer through horizontally inhomogeneous precipitating clouds. *J. Geophys. Res.*, **99**, D8, 16,707–16,718.

Roberti, L. and Kummerow, C., 1999: Monte Carlo calculations of polarized microwave radiation emerging from cloud structures. *J. Geophys. Res.*, **104**, D2, 2093–2104.

Roche, A. E., Kumer, J. B., Mergenthaler, J. L., Ely, G. A., Uplinger, W. G., Potter, J. F., James, T. C. and Sterritt, L. W., 1993: The Cryogenic Limb Array Etalon Spectrometer CLAES on UARS: Experiment Description and Performance. *J. Geophys. Res.*, **98**, D6, 10,763–10,775.

Rothman, L. S., Rinsland, C. P., Goldman, A., Massie, S. T., Edwards, D. P., Flaud, J.-M., Perrin, A., Camy-Peyret, C., Dana, V., Mandin, J.-Y., Schroeder, J., McCann, A., Gamache, R. R., Wattson, R. B., Yoshino, K., Chance, K. V., Jucks, K. W., Brown, L. R., Nemtchinov, V. and Varanasi, P., 1998: The HITRAN Molecular Spectroscopic Database and HAWKS (HITRAN Atmospheric Workstation): 1996 edition. *J. Quant. Spectrosc. Radiat. Transfer*, **60**, 665–710.

Rozanov, V. V., Diebel, D., Spurr, R. J. D. and Burrows, J. P., 1997: GOMETRAN: A radiative transfer model for the satellite project GOME - the plane-parallel version. . *J. Geophys. Res. 102 (d14) , 16683-16695.*

Schulz, F. M. and Stamnes, K., 2000: Angular distribution of the Stokes vector in a plane-parallel, verically inhomogeneous medium in the vector discrete ordinate radiative transfer. *J. Quant. Spectrosc. Radiat. Transfer*, **65**, 609–620.

Schulz, F. M., Stamnes, K. and Wenig, F., 1999: An improved and generalized Discrete Ordinate Method for polarized (vector) radiative transfer. *J. Quant. Spectrosc. Radiat. Transfer*, **61**, 1, 105–122.

Simmer, C., 1993: Handbuch zu MWMOD, Kieler Strahlungstransportmodell für den Mikrowellenbereich. Institut für Meereskunde Kiel.

Skofronik-Jackson, G. M., Gasiewski, A. J. and Wang, J. R., 2002: Influence of Microphysical Cloud Parametrizations on Microwave brightness Temperatures. *IEEE Transactions on Geoscience and Remote Sensing*, **40**, 1.

Spang, R., Riese, M., Eidmann, G., Offermann, D. and Wang, P. H., 2001: A detection method for cirrus clouds using CRISTA 1 and 2 measurements. *Adv. Space Res.*, **27**, 10, 1629–1634.

Sreerekha, T. R., Buehler, S. A. and Emde, C., 2002: A Simple New Radiative Transfer Model for Simulating the Effect of Cirrus Clouds in the Microwave Spectral Region. *J. Quant. Spectrosc. Radiat. Transfer*, **75**, 611–624.

Stallman, R. M., 2002: *Free Software, Free Society: Selected Essays.* GNU Press, ISBN 1-882114-98-1.

Stamnes, K., Tsay, S.-C., Wiscombe, W. and Jayaweera, K., 1988: Numerically stable algorithm for discrete-ordinate-method radiative transfer in multiple scattering and emitting layered media. *Appl. Opt.*, **27**, 2502–2509.

Stiller, G. P. (Ed.), 2000: *The Karlsruhe Optimized and Precise Radiative Transfer Algorithm (KOPRA)*, vol. FZKA 6487 of *Wissenschaftliche Berichte.* Forschungszentrum Karlsruhe.

Stiller, G. P., von Clarmann, T., Funke, B., Glatthor, N., Hase, F., Höpfner, M. and Linden, A., 2002: Sensitivity of trace gas abundances retrievals from infrared limb emission spectra to simplifying approximations in radiative transfer modeling. *J. Quant. Spectrosc. Radiat. Transfer*, **72**, 249–280.

Stubenrauch, C. J., Holz, R., Chedin, A., Mitchell, D. L. and Baran, A. J., 1999: Retrieval of cirrus ice crystal sizes from 8.3 and 11.1 mm emissivities determined by the improbed initialization inversion of TIROS-N Operational Vertical Sounder observations. *J. Geophys. Res.*, **104**, D24, 31 793–31 808.

Tjemkes, S. A., Patterson, T., Rizzi, R., Shephard, M. W., Clough, S. A., Matricardi, M., Haigh, J. D., Höpfner, M., Payan, S., Trotsenko, A., Scott, N., Rayer, P., Taylor, J. P., Clerbaux, C., Strow, L. L., DeSouza-Machado, S., Tobin, D. and Knuteson, R., 2003: The ISSWG line–by–line inter–comparison experiment. *J. Quant. Spectrosc. Radiat. Transfer*, **77**, 4, 433–453.

Toon, O. B., Tolbert, M. A., Middlebrook, A. M. and Jordan, J., 1994: Infrared optical constants of H_2O, ice, amorphous nitric acid solutions, and nitric acid hydrates. *J. Geophys. Res.*, **99**, D12, 25 631–25 654.

van de Hulst, H. C., 1957: *Light Scattering by Small particles.* New York: Dover Publications. Corrected republication 1981.

Wang, P.-H., Minnis, P., McCormick, M. P., Kent, G. S. and Skeens,

K. M., 1996: A 6-year climatology of cloud occurence frequency from Sratospheric Aerosol and Gas Experiment II observations (1985–1990). *J. Geophys. Res.*, **101**, 29, 407–429.

Warren, S., 1984: Optical Constants of Ice from the Ultraviolet to the Microwave. *Appl. Opt.*, **23**, 1206–1225.

Waterman, P. C., 1965: Matrix formulation of electromagnetic scattering. *Proc. IEEE*, **53**, 805–812.

Waters, J. W., Read, W. G., Froidevaux, L., Jarnot, R. F., Cofield, R. E., Flower, D. A., Lau, G. K., Pickett, H. M., Santee, M. L., Wu, D. L., Boyles, M. A., Burke, J. R., Lay, R. R., Loo, M. S., Livesey, N. J., Lungu, T. A., Manney, G. L., Nakamura, L. L., Perun, V. S., Ridenoure, B. P., Shippony, Z., Siegel, P. H., Thurstans, R. P., Harwood, R. S., Pumphrey, H. C. and Filipiak, M. J., 1999: The UARS and EOS Microwave Limb Sounder Experiments. *J. Atmos. Sci.*, **56**, 194–218.

Weng, F., 1992: A multi-layer discrete-ordinate method for vector radiative transfer in a vertically-inhomogeneous, emitting and scattering atmosphere – I. Theory. *J. Quant. Spectrosc. Radiat. Transfer*, **47**, 19–33.